GENETICALLY MODIFIED ORGANISMS

The Mystery Unraveled

M IRIAM J UMBA , P h D

Printed in Victoria, BC, Canada.

ISBN: 978-1-4269-1600-7 (sc)
ISBN: 978-1-4269-1601-4 (dj)

Library of Congress Control Number: 2009934817

*Our mission is to efficiently provide the world's finest, most comprehensive book publishing
service, enabling every author to experience success. To find out how to publish your book, your
way, and have it available worldwide, visit us online at www.trafford.com*

Trafford rev. 4/6/2010

Trafford
PUBLISHING® www.trafford.com

North America & international
toll-free: 1 888 232 4444 (USA & Canada)
phone: 250 383 6864 ♦ fax: 812 355 4082

DEDICATION

This book is dedicated to my children for giving me hope and inspiration. Special dedication goes to my brothers and sisters for tirelessly egging me on.

ACKNOWLEDGEMENTS

To my LORD and savior, thank you for your wisdom, and great inspiration that literally served to put the wording of this book into my mind.

For his partial assistance in reviewing the topic that appears in Chapter IV, titled: *Transgenic Mosquitoes Impaired in Transmission of Disease Pathogens*, I gratefully acknowledge Dr. Richard W. Mukabana of University of Nairobi, School of Biological Sciences.

A million thanks to my family for their moral and technical support. They were there for me when I needed them most, especially when I was faced with the challenges of writing this book.

To the Trafford Book Publishing team: Review Manager, Print Manager, Proofing Manager, Art Manager, and Press Manager; for their great effort in bringing the publication of this book to fruition.

Contents

PREFACE

It is a common observation that almost everyone you meet has a strong opinion about GM foods. Most concerns about GM foods fall into 3 categories: environmental hazards, human health risks, and economic concerns. Although, reports show that a lot of debate swirling around genetically modified crops or organisms, principally, is meant to mislead and confuse.

Opponents of genetically modified (GM) food have turned it into a specter, often referring to GM food as "Frankenfood," after Mary Shelley's character Frankenstein and the monster he creates in her novel of the same name. One author notes that "the term Frankenfood was coined in "1992" by Paul Lewis, an English professor at Boston College; who used the word in a letter he wrote to the *New York Times* in response to the decision of the US Food and Drug Administration to allow commercial companies to market genetically modified food."

Well, Frankenstein or not, genetically-modified foods have the potential to solve many of the world's hunger and malnutrition problems, and to help protect and preserve the environment by increasing yield and reducing reliance upon chemical pesticides and herbicides. Many people feel like I do, that genetic engineering is the inevitable wave of the future and that we cannot afford to ignore a technology that has such enormous potential benefits. Yet there are many challenges ahead for governments, especially in the areas of safety testing, regulation, international policy, and food labeling. However, if you go deep into the story of genetic modification, you will find that the benefits or rather advantages of these so called "new" products far outweigh the risks and concerns.

In support of GMOs, it is further argued that with the increasingly limited amount of new land available to agriculture, modern biotechnologies such as gene manipulation could complement and improve the efficiency of traditional selection and breeding techniques, to enhance agricultural productivity.

Virtually all types of foods and organisms have been genetically engineered: corn, cotton, tomatoes, soybeans, sugar beets, oilseed rape, maize, salmon, pigs, cows etc. - the list goes on. It is widely acknowledged that with about 6 billion people eating everyday, we need every reasonable tool known to man to ensure adequate nutrition for Earth's residents. GM foods, properly utilized, can help meet these needs in a number of ways: pest resistance, herbicide tolerance, disease resistance, cold tolerance, drought tolerance and salinity tolerance, among others.

Well, GMOs have certainly arrived! It is common knowledge, that with increased cross border interactions, and lack of stringent regulatory measures, GMOs have already found their way into Africa and other countries outside the African continent, that are opposers of GMOs. Therefore my argument is that since we are already partaking of these so called "new" products, our legislators might as well legalize their adoption. The Cartagena Protocol on Bio-safety requires public awareness concerning the safe transfer, handling and use of GMOs. However, most people from developing countries, especially Africa lack awareness on GMOs. It is therefore hoped that this review will serve to bridge the information gap.

Most African countries are yet to fully benefit from this new technology because of suspicion and ignorance of its potential. Survey studies carried out world wide indicate that there are a high percentage of people willing to embrace GMOs. Many developing countries are poor, and thus should not reject that which can contribute to the alleviation of poverty. For example, Africa's available land resources cannot keep pace with its rate of population increase. Available arable land is limited and family land holdings are getting smaller by the day. Recurrent droughts, rising food prices, famines and conflicts have intensified poverty. Therefore investment in GMOs or biotechnology can reduce our dependence on the west, and bring about positive development.

CHAPTER I

Introduction

The acronym GMOs stands for genetically modified organisms, and is also referred to as genetically engineered organisms (GEOs). A GMO is an organism whose genetic make-up has been deliberately altered. The GMOs are created when genes are transferred from one organism to another, to create a new type of unnatural organism. Resultant genetically engineered plants are called transgenic plants, while genetically engineered animals are called transgenic animals.

The genetic material is altered using various techniques in genetics. The following terminologies commonly refer to techniques employed in the manipulation of genes: (a) Recombinant DNA technology (b) Genetic engineering (c) Genetic modification (d) Gene splicing (e) Gene technology. These methodologies are all simply referred to as biotechnology.

It is a common observation that while most industries use mechanical devices - machines - to make things; biotechnology uses living organisms to make products of economic value; this is the basis of GMOs. One author notes that, the first known use of the word "biotechnology" was in 1919 by a Hungarian agricultural engineer, Károli (Karl) Ereky (the founding father of biotechnology). He used it to describe a system for raising pigs on sugar beets as their primary food source. Biotechnology was defined by Ereky as "all lines of work by which products are produced from raw materials with the aid of living things". [1, 2, 3, 6, 7]

Using biotechnology to Influence natural processes is hardly a novel concept. For thousands of years, scientists have conducted selected breeding to improve livestock; and controlled plant pollinators to produce better crops. [3, 6, 7]

Apart from the traditional cultivation of crops and raising of livestock, there are a number of other prehistoric and ancient technologies that can also be characterized as biotechnology. These include industrial applications such as the use of other microorganisms to make cheese, yogurt, pickles, some sausages etc.; and fermentation: brewing beer, wine-making, baking bread. Most of these food processing systems rely on yeast to alter the properties of a raw material. In the case of brewing and wine-making, yeasts convert the sugars in grains and grapes into alcohol. In bread, yeasts produce carbon dioxide to make the bread rise and soften the dough. [1, 3]

A simple process like sewage treatment also encompasses biotechnology. Sewage treatment is the processing of toxic domestic and industrial waste into less harmful materials known as sludge. In this process organic wastes are degraded by the action of a complex community of microbes. [1, 2, 3, 7] Some of the most important modern biotechnologies led to the development of vaccines and antibiotics. Fermentation systems were developed to grow microbes, which were then used for the production of antibiotics. However, large scale production of antibiotics was not possible until suitable fermentation systems were developed, and strains with higher yield of antibiotics were identified. In recent years, however, the scope of biotechnology has been altered beyond this original vision by a series of developments in the life sciences. Scientists can now identify and isolate individual genes responsible for producing specific characteristics. By altering these genes, or transferring them to another organism, the scientists can "engineer" the genetic makeup of that organism producing a new variety. This is the heart of the "biological revolution" that is going on at the moment. [1, 3, 4]

Recent achievements in gene manipulation have necessitated, or rather, created a need to re-define the word biotechnology. According to reports including the U.S. Office of Technology Assessment, biotechnology has recently been defined as "**Any technique that uses living organisms to make or modify products, to improve plants or animals, or to develop microorganisms for specific purposes.**" [3, 6, 7]

It is widely acknowledged that since the earliest days of agriculture, farmers have taken advantage of genetic differences between plants. For example, those varieties of corn and wheat that were more resistant to disease, pests or temperature variation, or which yielded more or better products were selected and cultivated. Conversley, those varieties that were susceptible to disease or demonstrated lower quality yields, were screened out. These preferred characteristics, namely disease resistance, high yield or quality, flower color etc. can be traced to gene expression. By selecting and cultivating plants with desirable characteristics, farmers caused certain desirable genes to be more abundant than the undesirable ones.

Until recently, the process of selecting plants with the optimum genetic characteristics required years of breeding and cross-breeding. However, through the use of newer biotechnology, scientists can now develop crops with desirable characteristics more quickly and less expensively by identifying the desired gene in another plant (or animal or microorganism), and integrating this desired gene into the recipient's plant genome, thus creating a transgenic plant or animal. Sometimes a gene from another plant of the same species is used, but most transgenic plants incorporate genes from other species. In addition, scientists can introduce genes that endow plants with characteristics that could never be achieved through cross-breeding, like the capability of producing vaccines or other drugs. Today, transgenic plants resistant to bacteria, viruses, insects, herbicides exist. [3, 5, 8]

Biotechnology-based developments are having a significant impact world-wide. Listed below are some of the industries that are being impacted by these new technologies:

1. Medicine - development of new drugs to cure and control human diseases. Notable examples are: protease inhibitors to treat HIV infection, and tissue plasminogen activator (tPA) to remove clots in blood vessels).

2. Diagnostics - new tests to rapidly and more accurately detect pathogens in medicine, the environment, food processing and agriculture.

3. Food products - a wide range of new food additives and processing reagents.

4. Environment - new methods to treat waste products, organisms to clean up environmental pollution (bioremediation), and new sources of energy.

5. Chemicals - production of feed-stocks for the chemical industry from renewable resources, enzymes for processing systems and other applications.

A general observation is that products derived through biotechnology [some of which are GMOs] have been relatively slow to reach the market. Today, some products of biotechnology revolution are available in the market, and many more are in the development pipeline. Examples of these products for agricultural use include: [5, 8]

- Tomatoes - with better flavor, longer shelf life, thicker paste;

- Corn and cotton plants with better insect resistance and less need for insecticides;

- Canola that produces better, healthier oils, and oils with specialized uses;

- Corn and soybeans that tolerate broad-spectrum herbicides such as RoundUp;

- Hormones produced in bacteria to increase milk production in cows;

- Enzymes for cheese making that are produced in bacteria, rather than from calves;

- Cows, sheep and goats that produce therapeutic drugs in their milk;

- Salmon that can grow 3 times faster than normal.

LITERATURE CITED

[1]Bud, R. (1993): The uses of life. A history of biotechnology. Cambridge University press. 299p.

[2]Ereky, K. (1919): Biotechnology der fleisch-, Fett-und Milcherzeugung im landwirtschaftii chen Grosbetriebe. Verlag Paul Parey, Berlin. 84P.

[3]Fári, Mi, Bud, R. and Kralovánsky, U. P. (2001): The founding father of biotechnology: Solving the Ereky enigma. Scientific American.

[4]Jaerisch Rudolf (1988). In: Science, Vol. 240, pages 1468–1474: June 10, 1988.

[5]Paleyanda Rekha, Janet young, William Velander and William drohin (1991). In: Recombinant Technology in Hemostasis and Thrombosis. Edited by L. W. Hoyer and W. N. Drohan. Plenum Press, 1991.

[6]Shmaefsky, Brian (2006). The definition of biotechnology. In his Biotechnology 101. Westport, CT, Greenwood press, 2006 p. 1–17.

[7]Smith, J. E. (2006). Public Perception of Biotechnology. Edited by Colin Ratledge and Bjorn Kristiansen. 3rd Ed. Cambridge, NY, Cambridge University Press, 2006. P. 3–33.

[8]Velander, H. William, Henryk Lubon & William N. drohan (1997). Transgenic livestock as drug factories. Scientific American January 1997, pages 70–74.

CHAPTER II

The Safety of GMOs

Ref.: [11, 13, 17, 20, 24, 30, 31, 33]

The big question on the lips of many lay people, or rather the million-dollar question is: "Are GMOs safe?" The salient facts regarding this new technology must be laid bare to pave way for resolutions that will lead to speedy adaptations to GMOs. That may come through legislative measures on a Global level. Many major controversies surround genetically engineered crops and foods. These commonly focus on the long-term health effects for anyone eating them. Some of these controversies are: environmental safety, labeling and consumer choice, intellectual property rights, ethics, food security, poverty reduction, and environmental conservation.

One of the controversial issues surrounding GMOs is the notion that seeds developed through this technology cannot be replanted after harvesting. This means the farmers must buy the "new" seeds every planting season. It is further argued that the cost of growing these so-called "new" crops will be prohibitive for peasant farmers due to the fact that there are specific pesticides to be used on GM crops. Then, there are those people who mistakenly think that this technology is dangerous - the fear that genetically modified food products are harmful and are aimed at reducing Africa's population. For some the concern is that it is unethical or immoral to tamper with the genetic material; they believe this will lead to a loss of our own "genetic privacy", and that no one has the right to own and patent either organisms or genes.

Others question the environmental consequences of these developments; that biotechnology will lead to the development of more potent pests, create new health problems because of allergens and toxins, and reduce biodiversity. Yet others question the socioeconomic consequences of biotechnology, for example the promotion of "corporate" farming, and the effects on the less developed countries [the belief that Western countries are taking advantage to offload their GMOs to our markets].

For the medical profession the row over genetically modified medicines is almost non-existent, as there is wide-spread acceptance and use of many recombinant products in health care. Recombinant products in health care include human insulin and growth hormone, erythropoietin, hepatitis B vaccine, tissue plasminogen activator, several interferons, factor VIII, and anti-haemophilic factor. It can obviously be seen that many people conveniently see medical applications of gene technology as "good genetics" but see genetically modified foods as "bad genetics."

Well, the irony of all this righteous indignation is that GMOs are the much needed solution; to alleviate hunger in Africa, Asia, and other developing countries. Droughts and famine are increasing throughout the world, particularly on the continent of Africa. In all certainty, starvation is much more dangerous to people than any threat presented by GM foods. It is therefore hoped that this review will provide information that will go towards demystifying the seemingly people's conception or rather misconception of GMOs, that has denied many a chance to embrace this new technology.

The issue of GMOs raises yet another pertinent question: "What is genetic modification and how does it relate to food production?" The following information review was given in answer to the above question, and is a passage extract from *British Medical Journal (1999) February 27; 318(7183): 581–584 Issue; by Leighton Jones, publications manager:* The author points out that, DNA is and always has been part of our daily diet. Daily, each of us consumes millions of copies of many thousands of genes. Many of these genes that we unknowingly ingest are fully viable at the point of consumption, and in most cases we do not know what they do. Have you ever stopped to consider the viable yet unknown genes of tomato, cucumber, and lettuce in a salad, the bovine genes in a beef steak, the fragmented DNA in many processed foods, and the genes of the many micro-organisms that we breathe and swallow?

A further observation made is that, genes change every day by natural mutation and recombination, thereby creating new species through the process of natural selection. Over the years, mankind has taken advantage of this natural variation, shuffling genes by selectively breeding wild plants and animals, and even microorganisms such as bacterial cultures used in making yogurt, and yeasts to produce domesticated variants that are better suited to his needs. Such selective breeding involves the transfer of unknown numbers and kinds of genes between individuals of the same species. Apart from such conventional means of gene transfer; other methods of gene transfer used before the advent of genetic modification technology included a kind of technology involving polyploidisation and mutagenesis via x–rays. For example, barley seeds (Golden promise) were treated with x–rays in a reactor to yield the UK's favorite variety of barley used in brewing. [19, 21]

Well, just about everything we eat is derived from livestock, crops, and microorganisms bred specifically to provide food. Humans have also redistributed genes geographically: The soybean is native to Asia but is now grown throughout the world; the potato is native to the American continent, but it is grown throughout the world. Notably then, DNA has never been "static," neither naturally or at the hand of people. Arguably, genetic modification, which is the basis of GMOs - is an extension of this. However, unlike conventional breeding, in which new assortment of genes are created more or less at random, it allows specific genes to be identified, isolated, copied, and introduced into other organisms in much more direct and controlled ways. [21]

The report further points out that the most obvious difference between this new technology and conventional breeding is that, genetic modification allows us to transfer genes between different species, for example engineering bacterial toxin genes into a plant. Genetic modification also allows individual genes to be specifically switched off or not to be expressed; for example, a tomato paste now commercially available in many Western countries is produced using this technology: The gene that controls fruit softening was selectively under-expressed (that is, turned down) in tomatoes. This gene codes for the enzyme polygalacturonase, which digests the pectin that cements the fruit cells together and acts as a natural thickener in tomato pastes. As less polygalacturonase is produced, more of the natural thickener remains in the ripe fruit; reducing the amount of energy required to thicken the paste.

Using this new technology of gene transfer, it is now possible to introduce foreign genes (transgenes) into crop plants and express these in specific tissues. For example the foreign genes can be expressed in specific plant tissues, such as the roots or leaves of the plant, and not in other tissues (such as seeds and fruits). This is likely to substantially improve crop protection. This technology can be used to develop transgenic plants with protection against pests which attack only certain parts of the plant, such as roots or leaves.

ADVANTAGES OF GENETIC ENGINEERING

There are several advantages of gene technology: Ref.: [8, 19, 20, 17, 36]

1. **Wider selection of traits.** Genetic modification (GM) allows a much wider selection of traits for improvement. For example, not only pest, disease and herbicide resistance; but also potentially drought resistance halo tolerance, improved nutritional content (yield and quality of specific nutrients e.g. enhancement of vitamin A, iron, or a specific amino acid; or removal of toxic components to a tuber such as cassava). It enables foreign genes (from outside species) to be introduced in an organism.

2. **It is faster.** GM is potentially faster and lower in cost than traditional plant breeding. It enables the transfer of specific genes as opposed to plant breeding.

3. GM reduces risk of random occurrence of undesirable traits such as dwarf varieties.

4. Arguably, almost 100 million people are expected to be added to the world's population each year for the next 30 years. It is only logical to surmise that without biotechnology, we won't be able to increase the availability of affordable basic food. Although biotechnology has potential risks, starvation is worse.

5. **Gene Therapy:** Gene therapy may be used for treating or even curing genetic and acquired diseases like cancer and AIDS by using normal genes to supplement or replace defective genes or to bolster a normal function such as immunity.

6. **Pharmaceuticals:** Medicines and vaccines often are costly to produce and sometimes require special storage conditions not readily available in third world countries.

Scientists can now produce genetically modified edible vaccines in crops, such as tomatoes and potatoes; but this approach is still experimental.

7. **Phytoremediation:** Not all GM plants are grown as crops. Soil and groundwater pollution continues to be a problem in all parts of the world. Plants such as poplar trees have been genetically engineered to clean up heavy metal pollution from contaminated soil.

8. Use of a growth hormone gene to produce genetically engineered fish that grow much faster than the wild.

9. Recombinant bovine growth hormone already enables cows to use feed more efficiently and produce more milk.

10. **Herbicide tolerance:** For some crops, it is not cost-effective to remove weeds by physical means such as tilling, so farmers will often spray large quantities of different herbicides (weed-killers) to destroy weeds. This is a time-consuming and expensive process which requires care so that the herbicide doesn't harm the crop plant or the environment. Crop plants genetically-engineered to be resistant to powerful herbicides could help prevent environmental damage by reducing the amount of herbicides needed.

   For example, reports show that Monsanto has created a strain of soybeans genetically modified to be not affected by their herbicide product Roundup. When a farmer grows these soybeans, he will only require one application of weed-killer instead of multiple applications, reducing production cost and limiting the dangers of agricultural waste run-off.

11. **Pest resistance:** many plant varieties can be engineered with inbuilt toxin production capability against pests. For

example transgenic crops have been developed by introducing toxins from a soil bacterium, *Bacillus thuringiensis* (Bt). Spores and crystalline insecticidal proteins produced by *B. thuringiensis* are used as specific insecticides under trade names such as Dipel and Thuricide. Because of their specificity, these pesticides are regarded as environmentally friendly, with little or no effect on humans, wildlife, pollinators, and most other beneficial insects. The Belgian company Plant Genetic Systems was the first company (in 1985) to develop genetically engineered tobacco plants with insect tolerance by expressing *cry* genes from *B. thuringiensis*. [16, 36] Currently, the main Bt crop being grown by small farmers in developing countries is cotton.

There are several advantages of expressing Bt toxins in transgenic Bt crops: The first is that the level of toxin expression can be very high thus ensuring sufficient dosage, as well as cutting down on the use of synthetic pesticides in the environment. Secondly, the toxin expression is contained within the plant system and hence only those insects that feed on the crop perish.

12. **Disease resistance:** There are many viruses, fungi and bacteria that cause plant diseases. Plant biologists are working to create plants with genetically-engineered resistance to these diseases.

13. **Cold tolerance:** Unexpected frost can destroy sensitive seedlings. An antifreeze gene from cold water fish has been introduced into plants such as tobacco and potato. With this antifreeze gene, these plants are able to tolerate cold temperatures that normally would kill unmodified seedlings.

14. **Drought or salinity tolerance:** As the world population grows and more land is utilized for housing instead of food production; farmers will need to grow crops in loca-

tions previously unsuited for plant cultivation. Creating plants that can withstand long periods of drought or high salt content in soil and groundwater will help people to grow crops in formerly inhospitable places.

15. **Nutrition:** Malnutrition is common in African countries which face hunger and poverty where impoverished peoples rely on a single crop such as rice as their staple diet. However, rice does not contain adequate amounts of all necessary nutrients to prevent malnutrition. Scientists argue that, if rice and other cereals could be genetically engineered to contain additional vitamins and minerals, nutrient deficiencies could be alleviated. For example, blindness due to vitamin A deficiency is a common problem in third world countries.

RISKS AND CONCERNS OF GENETIC ENGINEERING

Ref.: [8, 13, 19, 26, 28, 30, 34, 40]

1. **Herbicide-resistance or insect resistance** genes could spread from engineered crops to wild relatives and create super-weeds and bugs that are especially difficult to control.

2. **Some bioengineered products could wipe out the major exports of some developing nations.** For example, a genetically altered bacterium that produces vanilla flavoring is under development. This could eliminate the market for vanilla beans, one of Madagascar's major agricultural products. Also, recombinant bovine growth hormone is too expensive for the small scale dairy farmer, and so cannot compete with big companies.

3. **Possible production of allergens or toxic proteins not active to the crop:** Introduction of a gene into a plant may create new or non-expressed allergens which can cause adverse reactions in the susceptible individuals. Some analysts argue that this is just but a myth; "that there is no evidence whatsoever, showing that GM foods in general are any different from "normal" foods in terms of toxicity or allergenic potential. Justifiably, many of the genes used to modify plants occur naturally in plants, or they are associated with the pathogens that infect plants, meaning humans have already been exposed to them."

    Nevertheless there were reports of an isolated case of a GM food causing an allergenic reaction. A Brazil-nut protein was introduced into soya bean to increase the content of an essential amino acid methionine [26, 40]. However, this new line of soya bean was never commercialized, owing to the allergenecity created by the novel protein expressed in the new line of soya bean. The present example shows that new traits may also give rise to special risks, and must be thoroughly investigated before use in animal or human nutrition, or before being released into the environment.

4. **Adverse effects on non-target organisms,** especially pollinators and biological control organisms.

5. **Gene pollution or gene flow** is the unwanted transfer of genes to other species. The big concern is that genes from genetically modified crops can be passed on to other plants resulting in environmental havoc. Although, plants can only be pollinated by closely related crops. Transgenic pollen and seeds can only disperse into other plants that are closely related. Therefore it is important to look very closely at the potential of GM crops to cross-pollinate other plants, especially weedy relatives. Fortunately, major food crops such as maize do not have weedy relatives. That is to say, maize plants cannot cross-fertilize with other plant species. However, not to be overlooked is the possibility of gene flow from transgenic maize to land races. Transgenic DNA from GM maize has been found introgressed into traditional maize varieties in Mexico. [22, 28] This may lead to loss of biodiversity, since diversity of food crops is considered essential for Global food security. However, this is countered by the fact that hundreds of new varieties of crops are produced every year by conventional breeding, and this increases biodiversity.

Gene splicing to date has focused on altering the nucleus, which is the brain of the cell, because it controls the plants' activities. This allows for the permanent and heritable expression of the particular trait engineered. When the foreign gene is introduced into the host plant via the nucleus (e.g. using *Agrobacterium tumefasciens*), the genetic alteration is transmitted to all plant tissues including the pollen, making it difficult to control the spread of the engineered plants. [15]

In an apparent move to curb the phenomenon of gene flow, genetic modification of plants is shifting towards strategies that impart transient expression of the trait. Transient expression includes transforming chloroplasts using ballistic (gene-gun), or other methods. All plants contain cell organelles, which include the nucleus and chloroplasts. Chloroplasts are abundant, and they do not spread their genetic changes to any other cells within the plant. Alternatively, the desired gene fragment can be inserted into the host cells using a plant viral pathogen such as the tobacco mosaic virus without the actual incorporation of the new genetic material into the plant chromosomes. [15]

6. **Lack of "right to know"** – farmers, consumers and other stakeholders have been kept in the dark over the biological and economic impacts of the GMOs.

7. **Antibiotic resistance:** According to, *WHO (1993)* [37]; *Advisory Committee on Novel Foods and Processes (1994)* [9]; *Nordic Working group on Food toxicology and Risk assessment (1996)* [25]; *Border and Norton, M (1998)* [10]; *Leighton Jones (1999)* [21]; *OECD (2000)* [27]: Studies have shown that the process of transferring genes from one plant to another is complex and imprecise. It has been found that only a small fraction of the plant cells targeted with the new gene will actually incorporate that gene into their own genome; this makes it difficult to identify altered cells to be used for propagation. Geneticists have found a way of circumventing this problem by linking an antibiotic resistance gene (as a marker) to the gene to be cloned. Only cells that have incorporated the new genes

will grow on the selective growth media used, containing the antibiotic while others fail to grow.

Many major controversies have arisen over the use of antibiotic resistance as a marker system for genes. In general, it has been shown that the antibiotics used in marker systems are not used for treating diseases. It is also further argued that the gene and its product (that is, the enzyme that inactivates the antibiotic and thus confers resistance), is usually destroyed during heat processing of the food material. It is however, apparent that there may be cause for some of the alarms that have been raised. The report by Leighton Jones (1999) highlights the fact that there have been two cases, in which clinically important antibiotics have been used.

According to the report, a "new" maize variety developed by Novartis contained a gene for ampicillin resistance, and a potato developed by Avebe contained a gene for amikacin resistance. As maize is often fed to livestock; a particular concern is that the antibiotic resistance will get into microbes in the animals' gut and so be passed on to humans via dairy or meat products. This could reduce the efficacy of antibiotics in treating human and animal diseases.

The same report also points out that as a consequence of these concerns; both these genetically modified crops are having difficulties gaining full regulatory approval. Many countries are on guard against being consumers of this controversial maize type; Norway has banned all genetically modified organisms containing antibiotic resistance genes. EU-member countries Austria and Luxembourg have also banned the import of Novartis maize.

It is however, encouraging to note: that in an apparent remedial move, the use of antibiotic markers in the pro-

duction of genetically modified plants is being phased
out and replaced by other markers such as herbicide
resistance. Other methods of transformation without
using antibiotic marker are becoming available, such as
the use of sperm to transport foreign DNA into ova. The
use of sperm is reported to have a relatively high rate
of success. It is also technically simple to carry out, and
has potential for transferring larger pieces of DNA, and
it is applicable to animals. [21]

8. **The fear of farmers' dependency on patented GM
   seeds and other products.** Farmers and other consumers of genetically modified products have to content with
   the Intellectual Property Rights which grants inventors
   monopolies in exchange for their socially valuable innovations. Currently, Monsanto Company is reported to be
   the world's leading producer of genetically engineered
   seed, holding 70% – 100% market share for various crops.
   Apart from the genetically modified seed, Monsanto is
   also involved in developing and marketing bovine growth
   hormone. Bovine somatotropin, is a synthetic hormone
   that is injected into cows to increase milk production[39]

There is a general trend of owners of patents trying to
protect them, to maintain a hands-on operation on their
innovations. If the farmers can independently produce
seeds then the patent owner stands to lose their trade.
The justification is that research is a way to put billions
of dollars spent on research back into the system. [23, 39]
This has necessitated the development of exclusive GM
seeds, and the enforcement of exclusive rights to GM
products. Reports show that in 1998, various firms in the
United States, which are either producers of GM seeds
or stakeholders were awarded a patent on a technique
that genetically disables a seeds' ability to germinate
when planted a second season – this patent was dubbed

"terminator technology." This patent covered GM cotton, tobacco, and potentially all cultivated crops. Terminator technology produces plants that have sterile seeds, so they do not flower or grow fruit after the initial planting, requiring farmers to purchase seed for every planting season.[14, 35, 38, 39]

The terminator technology has met with so much opposition that it has not been commercialized to date. Monsanto is believed to have 87 terminator patents pending in developing countries but has never used the technology commercially, or even tested it in field trials. The technology is still in its infancy and most large GM companies are developing their own versions. The pressure to drop terminator-type technologies is still on. In recent years, widespread opposition from environmental organizations and farmer associations has grown, mainly out of concerns that these seeds increase farmers' dependency on seed suppliers especially in the developing countries. [14, 35, 38, 39]

A primary objective of 'terminator technology' is to grant and protect corporate rights to charge fees for patents on products that are genetically modified. This technology offers no advantage by itself, but when coupled with the production of the strongest, highest yielding seeds, farmers may be compelled to buy single-season plants. [14, 23, 35, 38, 39]

The concern raised by many analysts is that "terminator plants, if legalized, will effectively constrict worldwide crop diversity by preventing farmers from engaging in seed selection and cross breeding that has, for thousands of years, given them the ability to adapt crops to local conditions. Crop uniformity increases vulnerability to pests and diseases, thus heightening the potential for mass famine." [14, 23, 35, 39]

SAFETY ASSESSMENT OF GENETICALLY MODIFIED FOODS

In many countries, the safety assessment of GM foods relies upon the concept of "substantial equivalence" that must be demonstrated between the GM food and its conventional food counterpart. GM foods are considered to be "substantially equivalent" to conventional foods when levels of nutrients, allergens, or naturally occurring toxins are not substantially different and there are no allergens or toxins detected.[32]

This way of assessing safety of GM food is not full proof. In some cases, it is not easy to separate engineered from non-engineered foods. For instance, many food items on supermarket shelves contain soybean, from canned soups to baby foods. America is one of the largest suppliers of soybean and some of their exported soybean may have been genetically modified for insect or herbicide resistance. However, all the soybeans are pooled; and it is possible although very expensive to separate the GMOs from the non-GMOs. It is also argued that, the soybeans will have been cooked during the preparation of certain food items, a process which denatures proteins, including the ones produced by the introduced gene(s).

A GM food may not be rejected or rather condemned based on an allergic response, because under normal circumstances, consumption of any food (conventional, organic or GM) may present some risk of hazard due to the presence of proteins or other naturally occurring chemicals that might cause allergies or other harmful effects. The most common allergy-causing foods are cow's milk, eggs, fish, shellfish, tree nuts, wheat, peanuts, and soybean. [12]

One author notes that, while traditional approaches to assessing food examine the effects of individual chemicals on test animal species, these methods are impractical for studying the safety of genetically modified foods. This is due to the presence of thousands of unique chemicals in foods and the inability of test laboratory animals to consume large amounts of specific food items. The same author goes on to conclude that, "The shortcomings of current techniques used in assessing GM food warrants the need to develop better diagnostic methods, such as mRNA fingerprinting, proteomics and secondary metabolite profiling." [18]

LITERATURE CITED

[9]Advisory Committee on Novel Foods & Processes.
Report on the use of antibiotic resistance markers in ge-
netically modified food organisms. London: ACNFP; 1994.

[10]Border, P; Norton, M. *Genetically modified foods—benefits*
and risks, regulation and public acceptance. London:
Parliamentary Office of Science and Technology; 1998.

[11]Boucher D (1999) *The Paradox of Plenty: Hunger in a Bountiful*
World. Food First, 342. ISBN 978-0935028713.

[12]Clydesdale, F. M. (1996). Allergenicity of foods produced by ge-
netic modification. Food Science and Nutrition 36: 1–186.
StarLink EPA (U. S. Environment Protection Agency)
2000. Assessment of scientific information concerning
Starlink corn. FIFRA Scientific Advisory Panel Meeting.
November 28. SAP Report No. 2000–06. Epa Scientific
Advisory Panel web site.

[13]Ewen WB Stanley, and Pusztai Arpad, *Health risks of geneti-*
cally modified foods The Lancet Medical Journal: Vol. 354,
Issue 9179 August 1999.

[14]GLOBAL PESTICIDE CAMPAIGNER and EARTH ISLAND
JOURNAL, Title: "Terminator Seeds Threaten an End to
Farming," Date: June 1998, Fall 1998, Authors: Hope Shand
and Pat Mooney.

[15]Goldstein, D. A. and J. A. Thomas (2004). Biopharmaceuticals
derived from genetically modified plants. QJM vol. 97 No.
11 ©Association of Physicians 2004.

[16]Hofte H, de Greve H, Seurinck J., et al (December, 1986).
"Structural and functional analysis of a cloned delta endotoxin of Bacillus thuringiensis Berliner 1715". Eur. J. Biochem.

161 (2):273–280; doi:10.1111/j.1432-1033.1986.tb10443.x. PMID 3023091. http://www.blackwell-synergy.com/openurl?genre =article&sid=nlm:pubmed&issn=0014-2956&date=1986&vo lume=161&issue=2&spage=273.

[17]Kok, E. J. & H. A. Kuiper (2003). Comparative safety assessment for biotech crops. Trends in Biotechnology 21 (10): 439–444.

[18]Kuiper, H.A., Noteborn, H.P.J.M. and Peijnenburg, A.A.C.M. (1999) Adequacy of methods for testing the safety of genetically modified foods. *The Lancet* 354, 1315–1316.

[19]Institute of Food Science and Technology (IFST, 2008). Genetic Modification and Food p. 1–44.

[20]Lachmann, A. (1999). GM Food debate. Lancet 354–69.

[21]Leighton Jones, publication manager: BMJ. 1999 February 27; 318 (7183): 581–584. Copyright © 1999, British Medical Journal - Science, medicine, and the future - Genetically modified foods.

[22]Mercer L. Kristin and Joel D. Wainwright (2007). Gene flow from transgenic maize to landraces in Mexico: An analysis: doi:10.1016/j.agee.2007.05.007

[23]MOJO WIRE Title: "A Seedy Business" http://www.mother-jones.com/news_wire /broydo.html http://www.mother-jones.corn/news_wire /usda-inc.html, Date: April 27, 1998, Author: Leora Broydo; THIRD WORLD RESURGENCE #92 Title: "New Patent Aims to Prevent Farmers From Saving Seed," Author: Chakravarthi Raghavan.

[24]Nesbitt CT: GEO-PIE Project: U.S. regulation of genetically engineered organisms. 2005. Accessed Nov. 11, 2008 http:// www.geo- pie.cornell.edu/educators/who.html.

[25]Nordic Working group on Food toxicology and Risk Assessment. S. Kärenlampi. Health effects of marker genes in genetically engineered food plants. 1996. Tema Nord 1996:530. Department of Biochemistry and bio-technology University of Kuopio, P.O Box 1627, FIN-70211 Kuopio, Finland.

[26]Nordlee, J.A., S.L. Taylor, J.A. Townsend, L.A. Thomas, and R.K. Bush. 1996. Identification of Brazil-nut allergen in trans-genic soybeans. N. Engl. J. Med. 334:688–692.

[27]OECD, Environment Directorate Joint Meeting of the Chemicals Committee and the Working Party on Chemicals. Working group on harmonization of regulatory oversight in biotech-nology, Draft response to the G8 2000.

[28]Pilson D., HR Prendeville (2004). Ecological Effects of trans-genic Crops and the Escape of transgenes into Wild Populations. Annual reviews of Ecology, Evolution and systematic.

[29]Quist David and Chapela H. Ignacio (2001). Transgenic DNA introgressed into traditional maize landraces in Oaxaca, Mexico. *Nature* 414, 541-543 (29 November 2001) | doi:10.1038/35107068; Received 26 July 2001; Accepted 31 October 2001.

[30]Rosegrant MW, and cline Chinese SA (2003). Global Food Security: Challenges and Policies. Science 302: 1917–1919.

[31]Ryder EJ: The New Salad Crop Revolution. *Trends in new crops and new uses* In: Janick J & Whipkey A (eds). 2002, 408–412.

[32]Schauzu, M. (2000). The concept of substantial equivalence in safety assessment of foods derived from genetically modified organisms: AgBiotechNet 2 (April) ABN 044. Bundesinstitut fur Risikobewertung Web site.

[33]Shah A (2002): GE Technologies will solve world hunger. Global Issues 2002, accessed November, 11, 2008:http://www.globalissues.org/EnvIssues/GEFood/Hunger.asp

[34]Snow, A. A., D. A. Andow, P. Gepts, E. M. Hallerman, A. Power, J. M. Tiedje, and L. L. Wolfenbarger. (2005). Genetically engineered organisms and the environment: Current status and recommendations. Ecological Applications, July 16, 2004.

[35]THE ECOLOGIST, Title: "Monsanto: A Checkered History," and "Revolving Doors: Monsanto and the Regulators," Date: September/October 1998, Vol. 28, No. 5, Author: Brian Tokar.

[8]Velander, H. William, Henryk Lubon & William N. Drohan (1997). Transgenic livestock as drug factories. Scientific American January 1997, pages 70–74.

[36]*Vaeck M, Reynaerts A, Hofte A, et al. (1987). "Transgenic plants protected from insect attack". Nature 328: 33–7: doi:10.1038/328033a0. http://www.nature.com/nature/journal/v328/n6125/abs/328033a0.html.*

[37]WHO Food Safety Unit. Health aspects of marker genes in genetically modified plants, 1993. Report of WHO Workshop.

[38]http://www.guardian.co.uk/science/1999/oct/06/gm.food2 World braced for terminator 2. The Guardian (1999). Retrieved on 28 January 2008.

[39]http://en.wikipedia.org/wiki/Monsanto - 2009

[40]http://www.cgiar.org/biotech/rep0100/Lehrer.pdf: 2009. Potential Health Risks of genetically Modified Organisms: How can Allergens be assessed? Sammuel B. Lehren.

CHAPTER III

Genetically Modified Medicines

It is a common observation that when it comes to the application of genetically engineered medicines, they are taken as a lifeline, that is to say, it is normally a choice of life and death. When people are ill and risk death, they want to recover no matter how, even through taking genetically engineered medicine.

It is widely acknowledged that some of the so-called 'new' medicines are manufactured by a process that uses a genetically modified organism. In the majority of these cases, a bacterium or yeast will be modified to enable it to produce a naturally occurring human protein or pharmaceutical protein. The resulting medicine (i.e., the protein with therapeutic properties) will not typically contain any DNA (modified or otherwise), and the protein will be chemically very similar or identical to that normally produced in humans. In other words, while the protein is produced through a process involving genetic modification, the protein itself is not genetically modified. Many of these types of medicines, such as insulin and human growth hormones, have been approved for use in many developed countries. However, some of the 'new' types of medicines, such as vaccines, may contain live genetically modified organisms. Therefore, there is an educational need to inform and educate would be users so that they can make more informed decisions regarding the adaptation and usage of genetically engineered medicines. [8, 34]

RECOMBINANT HUMAN INSULIN

One author notes that, the first "new biotechnology" product that went to market was human insulin for the treatment of diabetes mellitus. "The first injection of insulin was given to a human diabetic patient on January 11, 1922. The patient had been declared terminal, and like all diabetics in those days, was drifting rapidly away. The response of the diabetic patient was found to be immediate and breathtaking. A man inches away from death recovered much of his old vitality, he ate his first meal in days, and his urine showed a complete absence of glucose." [8, 34, 43, 49]

Diabetics are unable to produce satisfactory amounts of insulin, which facilitates the processing of sugars from food into energy that the body can use. In the past, diabetics needed to take insulin purified from pigs and cows to fulfill their insulin requirement. However, bovine or porcine insulin are not quite identical to human insulin. In some patients, the pig or cow protein was recognized as foreign by the immune system.

Recombinant DNA technology has provided a way for scientists to produce human insulin in the laboratory. The gene for human insulin is isolated from human cells and inserted into plasmids. These plasmids are then introduced into bacterial cells which manufacture the insulin protein based on the human code. The purified product is identical in nature to human insulin and does not cause any allergies. In addition to use of bacteria to manufacture human insulin, other biotechnologists use yeast in recombinant DNA technology to make the product. Yeast can perform more of the complicated cellular processes that occur in human cells, making it a more useful organism for producing human substances. [8, 34, 43, 44]

DESIGNER BACTERIA AND VIRUSES TO FIGHT CANCER

Viruses are usually associated with disease. In recent years, however, scientists have turned a foe into a potential good by taking advantage of the ability of viruses to infiltrate human cells, where they can then reproduce. Viruses have developed this ability over the course of millions of years of evolution, and they are especially agile pirates, commandeering a cell's own machinery to churn out more viruses, and ultimately destroying the cell.

Reports show that researchers have genetically engineered certain virus strains (e.g. Herpes simplex virus-1) so that they will not cause disease. This has been done by switching off a gene that normally enables the virus to multiply and kill healthy cells. However, this so-called "new or designer" virus is able to infect rapidly dividing cancer cells. As a result the virus replicates inside tumor cells and destroys them without affecting healthy cells. [43, 47, 63]

In addition to using genetically altered viruses to treat cancer, in an analogous way, scientists have also incorporated the use of a bacterium (*Salmonella*) to treat cancer. *Salmonella* is commonly known for its role in causing food poisoning and septic shock. The "wild type" (unaltered form) of salmonella is genetically modified and stripped of its ability to cause illness. Once stripped, the bacteria have been successfully used to shrink solid tumors in laboratory animals. The modified form also inhibits the growth of the tumors as well. The Salmonella could thus be used to treat cancer in a place where the bacteria can thrive, like the colon and intestines. [45, 47, 63]

The genetically engineered *Salmonella* is reported to treat cancer in two ways: First is by naturally strengthening the human immune system. Secondly, the genetically altered *Salmonella* holds Interleukin-2 (Interleukin-2 is currently used in treating aggressive cancer such as melanoma), a naturally occurring cancer fighter in the body. The designer *Salmonella* release the drug to trigger a receptor in cancer cells called the "death receptor," which induces cancer cells to kill themselves. [79]

The genetic make-up of the designer *Salmonella* has been altered to give the cells a very short life-span. This is meant to ensure that once the Salmonella has administered the Interleucin-2, the bacterium does not linger in the patients' body for too long. Trial treatment using the combined effect of the bacterial enhancement of the immune system and the introduction of Interleukin-2 was found to be successful in lab animals. Reports show that a lot of progress has been made on this line of research. The first phase of human trials has started way back in 1999, and the Salmonella is currently being administered to patients. [41, 82, 83]

BLOOD CLOTTING OR COAGULATION

Hemophilia is an inherited disease attributed to lack of a blood-clotting factor.   Blood clotting is the same as coagulation. Most of the proteins involved are called Factor something, where "something" is a Roman numeral. Primarily, hemophilia is treated by replacing the absent or abnormal clotting factors to prevent severe blood loss and complications from bleeding. Clotting factors are replaced by injecting (infusing) a clotting factor replacement into the veins. Infusions of clotting factors help blood to clot normally and prevent damage to muscle, joints, and organs. [58, 70]

In the past, in order to get enough clotting factors, a person with hemophilia had to be injected with blood plasma. However, the problem with this is that clotting Factors purified from human blood were contaminated with some viral pathogens. Reports show that the clotting Factors were found to be contaminated with the Human Immunodeficiency Virus (HIV) and hepatitis B virus (HBV). Many hemophiliacs became victims of HIV infection through this route of transmission. [58, 70]

An alternative treatment comes in form of lab-produced (recombinant) factor concentrates that do not contain any human or animal products, so there is almost no risk of contracting a virus. Recombinant clotting Factors are currently used as therapy for hemophilia (esp. Factor XIII for treatment of hemophilia A and Factor IX for hemophilia B). The recombinant factor VIII concentrate was first introduced in 1987 to treat hemophilia A patients, and the product was licensed in the United States in 1992. [58, 70]

APPLICATION OF GENE THERAPY IN THE TREATMENT OF Alzheimer's Disease

A number of different diseases are characterized by the deposition of abnormal protein deposits in various organs, e.g. kidney, spleen, liver or brain. These 'amyloid' deposits consist of accumulations of various proteins in the form of plaques or fibrils depending on their origin, e.g. Alzheimer's disease; and variant Creutzfeldt-Jakob disease (vCJD), the human version of mad cow disease, is characterized by the deposition of plaques composed of beta-amyloid protein. Studies have shown that beta-amyloids compounds can be decomposed by the enzyme neprilysin, a naturally occurring protein that stops cell death and stimulates cell function. But this enzyme occurs much less often in brain areas where beta-amyloid has accumulated. [57, 60, 74]

A common approach at gene remedy of Alzheimer's disease involves the introduction of genetically modified cells or tissue containing the gene for neprilysin enzyme into the brains of Alzheimer's patients, using a virus vector. On an experimental level in the animals, the therapy was found to restore old, shrinking brain cells back to near-normal size and quantity, as well as connections essential for communication between the cells. An alternative approach at gene therapy of Alzheimer's disease, involves using genetically modified cells containing a gene for nerve growth factor (NGF). Research has shown that NGF can protect the type of neurons that die in Alzheimer's disease, and seems to encourage the remaining cells to proliferate and work more effectively. [57, 60, 74]

Currently, a genetically modified virus vector is being employed to infect the cell or tissue with the NGF gene, or the neprilysin gene. Researchers at the Board of Governors Gene Therapeutics Research Institute at Cedars-Sinai Medical Center demonstrated in an animal model that the delivery system for the gene, a novel gutted adenoviral vector called HC-Adv, is completely invisible to the immune system. Reports show that vectors previously used to deliver genes carried minute amounts of viral proteins that were detected by the immune system, triggering an immune response that rendered the therapeutic gene inactive after a period of weeks. Using these new technologies, gene therapy has successfully slowed the progression of Alzheimer's disease in infected patients. Gene therapy has also been successfully used to correct many other genetic defects, or human diseases known to be due to single gene defects, such as severe combined immune deficiency. [57, 59, 74]

Gene therapy has also found other applications. It has been used to correct a genetic defect known as Bubble baby that occurs due to a single gene defect-adenosine deaminase; Lesh-Nyhan-Severe mental retardation, a self-destructive behavior due to one brain enzyme missing; and Cystic fibrosis-rectified by inhaling DNA with normal gene.

RECOMBINANT TISSUE PLASMINOGEN ACTIVATOR (tPA)

Blood clotting is essential, for it prevents a disease condition known as hemophilia. However, a blood clot left in the blood circulation (esp. after surgery) can be fatal; thus it is also essential to be able to dissolve the clot eventually. Tissue plasminogen Activator (tPA) is a recombinant human protein which is chemically identical to endogenous tissue plasminogen activator. tPA activates a whole cascade of processes that break down blood clots. [48, 76, 80]

Heart attack or what is commonly referred to as stroke is a medical emergency. Previous studies indicate that the administration of recombinant tissue plasminogen activator (t-PA) to dissolve the clot improves the outcome after stroke when given very early to the stroke victim. A news brief report shows that the European Union recently approved its first GM medicine, Atryn, an anti-blood clotting medicine produced in the United States. [80]

GENETICALLY MODIFIED BACTERIA FOR THE PREVENTION OF TOOTH DECAY

A bacterium that is known to cause tooth decay (*Streptococcus mutans*) has been genetically altered into a harmless form that may permanently prevent the disease. *Streptococcus mutans* is a naturally occurring bacterium found in the mouth. This bacterium adheres to and grows on the teeth as a biofilm community, and it breaks down food sugars, resulting in the formation of lactic acid. Over time the lactic acid destroys tooth enamel, causing cavities. [78]

Recent reports show that this bacterium has been genetically altered into a friendlier version by removing the gene responsible for lactic acid production. "The new stain does not produce lactic acid and therefore will not cause tooth decay." This new or designer bacterium has been shown to tolerate high sugar levels without causing tooth decay in experimental animals. "Treatment using this bacterium will involve squirting a liquid solution of the effecter strain on the patients' teeth. The ideal application is said to be to treat infants when their first teeth appear." [78]

GENETICALLY MODIFIED PLANT-BIOPHARMACEUTICALS

Through genetic engineering, plants can now be used to produce pharmacologically active proteins, including human/ mammalian antibodies, human blood product substitutes such as plasma (plants producing blood), blood factors, anticancer drugs, vaccines, hormones and a variety of other therapeutic agents. [15, 46, 54, 53, 77]

There are a number of gene expression strategies that can be used to produce specific proteins in plants. One such strategy is referred to as transient expression (TE). With TE, the desired gene fragment can be inserted into the host cells using a plant viral pathogen such as the tobacco mosaic virus (TMV), ballistic (gene-gun), or other methods, without the actual incorporation of the new genetic material into the plant chromosomes. [15, 54, 53, 77] A biotechnological firm has exploited the capability of TMV as a vector. Selected genes involved in the production of blood plasma have been introduced in the TMV. Subsequently, the TMV was made to infect tobacco plants, turning them into a protein factory producing plasma (tobacco plants producing human blood), human growth hormone somatrophin, or anticancer drugs interleukin-2 or interferon. [46, 54, 53, 77] However, TE systems have one big disadvantage: the foreign trait does not get passed on to daughter cells. Thus TE systems require fresh production of transformed plants with each planting, and are not cost-effective for long term or high-volume protein production. [15]

Alternately, the plant chromosome can be altered to allow for the permanent and heritable expression of a particular protein, i.e. allow the creation of plants which produce seed carrying the desired modification. This can be done using the bacterium, *Agrobacterium tumefasciens*, a pathogen of plants that naturally infects plants and transfers its' genetic material to the plant chromosome. The desired protein can be selectively produced in seed or other tissues of the plant, using selective promoter systems. [15]

Plant-derived proteins have several advantages: (a) genetically engineered plants, acting as bioreactors, can efficiently produce recombinant proteins in larger quantities than those produced using mammalian cell systems; (b) plant-derived proteins are particularly attractive, since they are free from human diseases and viruses; (c) large quantities of biomass can easily be grown in the field, and may permit storage of material prior to processing. [15, 46]

HIV-T-CELL BASED VACCINE

Arguably, the Human Immunodeficiency virus (HIV) originated from wild chimpanzees and other wild animals in West Africa (Cameroon). A virus SIVcpz (Simian Immunodeficiency Virus) from chimps was shown to be the cause of HIV in humans. Through mutations, the animal version of HIV, that is SIV entered humans and mutated to become HIV. Scientists believe this transfer occurred more than once because of the multiple HIV strains infecting humans. People hunting chimpanzees and other animals first contracted the virus in the DRC in 1930. But it took 50 yrs before the virus was named due to varied symptoms in individuals and rareness of cases. It is estimated that, the human immunodeficiency virus that causes AIDS infects 33 million people globally, with approximately 2.7 million new infections reported each year, mostly in Africa. HIV belongs to a group of viruses called retroviruses. They are termed retroviruses because they contain an enzyme that transcribes DNA from an RNA template which is known as reverse transcriptase. [81, 84, 85]

The virus replicates itself rapidly, once it enters the body; HIV's power stems from its ability to mutate rapidly to evade detection and destruction by the body's defense mechanisms. The virus attacks a person's natural defense against disease (the immune system), weakening it over time. In particular HIV attacks one type of immune cell, known as CD4 T cells or lymphocytes, which help mount a defense. The T cells are components of the immune system that attack and destroy cells within the body that are infected. The HIV virus can also disguise itself to escape CD8 killer cells, also known as cytotoxic T lymphocytes or CTLs. The "CTLs are therefore crucial for the control of HIV infection. Some researchers have observed that unfortunately, HIV has an arsenal of mutational and non-mutational strategies that aid it in circumventing the CTL response mounted against it by its host. [51, 71, 81, 84, 85]

Recent studies carried out on genetically engineered immune cells have rekindled hope of finding a cure for AIDS. Researchers have developed a new killer T-cells, dubbed "assassin cells" therapy for treating HIV. The technology involves engineering the patient's own immune system to fight the virus more effectively. It is reported, that the researchers at the University of Pennsylvania took T-cells from an HIV patient and created a genetically engineered version that recognizes this deception. T cells are components of the immune system that attack and destroy cells within the body that are infected. In the patient, the T cell receptor protein seemed particularly good at recognizing HIV antigens. [54, 65, 68, 69, 81]

The research team isolated the receptor protein and then improved its ability to recognize HIV further by randomly mutating it. The killer T-cells were able to recognize other cells infected by HIV and slow the spread of the virus in in-vitro studies in laboratory dishes. The lead researcher projects that, "Billions of these anti-HIV warriors can be generated within a very short period, such as two weeks." [79]

Other reports show that, this innovation may be flawless: not only could the engineered T-cells see the stray HIV strains that may escape detection by natural T-cells, but also the engineered T cells were observed to respond in a much more vigorous fashion so that far fewer T-cells are required to control infection. Treating patients will involve taking a blood sample and adding an engineered virus containing genes for the improved T cell receptor. The patient's own T cells then take up the genes thereby equipping themselves with the improved receptor. These cells are then injected back into the patient. Through laboratory tests, researchers have found that killer cells given a new version of the natural T cell receptor are able to recognize all versions of a key HIV 'fingerprint' on the surface of the infected cell and subsequently clear HIV infection. The findings of the study reported in Nature Medicine may have important implications for developing new treatments for HIV and slowing or even preventing the onset of AIDS. The success of this 'assassin cell' therapy, which has proved effective in laboratory tests using human cell cultures, is projected to be tested in a clinical trial of HIV patients with advanced HIV infection that is due to start sometime in 2009. [68, 81]

In yet another technological development, an experimental AIDS vaccine has been found to prevent infection in monkeys. The researchers used the animal version of HIV, known as simian immunodeficiency virus (SIV), to simulate a model vaccine that can be used to treat HIV. The SIV virus used contained a single SIV protein that is able to prompt the animals to produce masses of "killer T-cells" that hunt and destroy SIV infected cells. One year after administering the first shots of the vaccine, the researchers are reported to have given the monkeys a lethal dose of SIV, and found that their T-cells reduced SIV viral levels by as much as 250-fold, thereby stopping the spread of the virus and also preventing the animals from progressing to simian-AIDS. Four of the six monkeys that received placebo shots are reported to have died after approximately 350 days. [68, 81]

RECOMBINANT (DNA) VACCINES

DNA vaccines are made of a modified form of an infectious organism's DNA that is introduced into plasmid vectors (derived from bacteria). The construction of bacterial plasmids with vaccine inserts is accomplished using recombinant DNA technology, as follows: The gene of interest is linked to a specific promoter or enhancer that allows protein expression in mammalian cells, resulting in the formation of a vaccine plasmid. Once constructed, the vaccine plasmid is transformed into bacteria, where bacterial growth produces multiple plasmid copies. The plasmid DNA is then purified from the bacteria by separating the circular plasmid from the much larger bacterial DNA and other bacterial impurities. This purified DNA acts as the vaccine. [42, 50, 52, 62, 73, 75]

Several researchers have shown that this material, when injected into muscle cells of a person, leads to the expression of the foreign gene. This gene expression ultimately leads to synthesis of infectious organism proteins inside the injected cells. As a result, the person's immune system responds in a protective manner by producing antibodies, which confer protection against the given pathogen. This happens almost in the same way as would occur if the person were actually infected by the true organism itself. Reports indicate that since the first report of a DNA vaccine was released, there has been an explosion of work on numerous pathogens with the hope of introducing a new era of immunization against diseases which have not yielded to conventional vaccine production techniques. DNA vaccines of some major diseases, including influenza, malaria, HIV, herpes simplex virus (HSV), colon cancer and cutaneous T cell lymphoma have been developed. However, most of these DNA vaccines are still at the early investigative stages and have not yet reached the level of being used in a human clinical trial. [40, 50, 52, 62, 73, 75]

An alternative method of administering some of the DNA vaccines is by spraying viral genes directly through the skin. This is a new technique that turns infinitesimal amounts of DNA into an effective vaccine. The technique involves coating microscopic particles with the vaccine; and subsequently shooting them into the human body at super-fast speeds using this new, needle-free device. Studies have shown that it is after the DNA gets into the cells of the skin that it produces such a strong immune response. If approved for use in humans, the new procedure could save lives in case of a flu pandemic, by skipping the conventional, time-consuming production of vaccines in chicken eggs. With current technology, DNA vaccines can be manufactured very rapidly and they can be manufactured in large amounts. [52, 75]

Advantages of DNA vaccines:

DNA immunization offers many advantages over the traditional forms of vaccination. Vaccines produced through recombinant DNA technology may be safer. For some diseases, live vaccines (which contain live viruses) are still used. There is a general concern that these may not always be as mild a strain of the virus as believed and may in some individuals cause a severe disease. If a protein from the virus can be expressed in microbes and used as a vaccine instead, it reduces the risk of unintentional infection. Researchers have observed that, DNA immunization is able to induce the expression of antigens that resemble native viral epitopes more closely than standard vaccines do. This is due to the fact that live attenuated and killed vaccines are often altered in their protein structure and antigenicity. [42, 50, 52, 62, 73, 75]

Another advantage of DNA vaccines is that the plasmid vectors used can be constructed and produced quickly and the coding sequence can be manipulated in many ways. DNA vaccines encoding several antigens or proteins can be delivered to the host in a single dose, only requiring a microgram of plasmids to induce immune responses. Rapid and large-scale production can be achieved at costs considerably lower than traditional vaccines. In addition, DNA vaccines are very temperature stable making storage and transport much easier. [42, 50, 52, 62, 73, 75]

Finally, genetic vaccines have therapeutic potential for ongoing chronic viral infections. DNA vaccination may provide an important tool for stimulating an immune response in patients with viral diseases, including HIV. Researchers believe that the continuous expression of the viral antigen caused by gene vaccination in an environment containing many APCs may promote successful therapeutic immune response which cannot be obtained by other traditional vaccines. [42, 50, 52, 62, 73, 75]

EDIBLE VACCINES

Recombinant vaccines offer a faster way of producing vaccines that are safe (free from contamination with animal viruses) and have the potential for being highly effective in preventing disease, both in humans and animals. The production of antigens in genetically-engineered plants could provide an inexpensive source of edible vaccines and antibodies to help in the fight against infectious diseases such as rabies, cholera, hepatitis B, malaria, and AIDS. But their popularity may be hindered by their high cost of production. Therefore, transgenic plants that express vaccine antigens represent an economical alternative of producing recombinant vaccines at more affordable rates. The transgenic material can be fed directly to man or animals as their oral dose of recombinant vaccine. This includes use of (whole or homogenized) leaves, fruits, or vegetable tissues. As an alternative route of administration, crude leaf extracts of the plant vaccine can be formulated and taken orally. [56, 59, 61, 64, 65, 67, 72]

This new technology of vaccine production could benefit third world countries which lack the infrastructure and resources to provide access to doctors. Specific vaccines have been produced in plants as a result of the transient or stable expression of foreign genes. It has recently been shown that genes encoding antigens of bacterial and viral pathogens can be expressed in plants in a form in which they retain native immunogenic properties. Transgenic potato tubers expressing a bacterial antigen have been shown to stimulate an immunological response in mice fed on these plants. These findings show that plants can be used as a vehicle to produce vaccines. [56, 59, 61, 64, 65, 67, 72]

The first test of edible vaccines was performed by expressing a surface protein of Hepatitis B in potatoes which were then fed to mice. The mice developed antibodies to the Hepatitis surface protein, and developed a mucosal immunity to infection by the virus. It is important to note that the antibodies are secreted by the mucosal membranes (lining of nose, mouth, digestive track) which is the site through which the virus is likely to invade the body. However, some researchers suggest that there may be many problems related to this new technology of vaccine production. For one, any particular protein might not be immunogenic. Ingesting too much protein could create a tolerance instead of an immune response. If vaccines are put into food prior to cooking, they may be denatured at the typical cooking temperatures, since many proteins aren't heat stable. In a bid to circumvent some of these problems, some researchers have used fresh edible plants such as bananas to produce vaccines. However, the problem with some of these plants especially bananas, is that they can only be grown under tropical conditions. A better alternative is to use tomatoes as factories to produce vaccines, because tomatoes can grow in many different climate zones and under varied conditions, and furthermore the tomato fruit can also be eaten fresh. [56, 59, 61, 64, 65, 67, 72]

THE TECHNOLOGY OF USING MICROBES AS FACTORIES IN THE PRODUCTION OF MEDICINE

Ref.: [8, 34]

The first transgenic organisms were bacteria – some of the simplest forms of life. Transgenic bacteria are still used today by pharmaceutical companies to produce a variety of human proteins such as insulin, human growth hormone and interferon, a family of substances with antiviral abilities. This offers a method to produce proteins that could not otherwise be purified in sufficient quantity for practical use. The ability to transform microbes (bacteria and yeasts) has made it possible to use these organisms to produce large amounts of specific proteins that are of high value for use in medicine, agriculture and other processing activities.

It is procedural that in order to synthesize large amounts of a protein in microbes; one needs to know the sequence of amino acids in the protein. This can be obtained directly by sequencing the protein, after which multiple copies of the gene can be obtained using polymerase chain reaction (PCR). The desired gene is then incorporated into the genome of the host microbe in a process known as transformation. The product of such a transformation is known as chimeric genes (chimeric genes are genes that are made of DNA of different organisms. The word chimeric is a translation of the Greek word chimera = mythological animal consisting of parts of different animals (lion body with bird head etc.).

The foreign gene that is introduced into the host organism will not be expressed unless it is attached to a promoter (a sort of biological switch) that will switch on the enzyme to produce the gene. The promoter can be made to be active in specific tissues of the human body (pancrease, skin etc.), or the plant (roots, anthers etc.). For example, a chimeric gene can be created from gene A and gene B as follows: Take a promoter from gene A by cutting it out using enzymes (restriction endonucleases) and ligate it with the DNA sequence that codes for enzyme B e.g. insulin production. Therefore these promoters are important because they can be used to synthesize proteins at the required site. Tests are then run to determine how much protein is produced by the transformed strain of microbe.

A fermentation procedure is developed to grow large volumes of transformed microbes. Again this must be optimized to obtain a high yield of the product. Conditions must be designed to allow expression of the target protein, usually depending on the promoter used to drive expression of the gene of interest. For example, some promoters are induced by environmental conditions such as high temperature. Others are induced by the addition of specific chemicals to the medium. In most cases, fermentation is allowed to proceed without synthesis of the target protein until the microbial cells have reached a high density close to a stationary growth phase. This is followed by induction of the promoter of the chimeric gene to initiate synthesis of the protein. The microbes are then harvested. The protein expressed in the microbes must then be purified from the cells. This is essential for a number of reasons. One is to remove bacterial proteins that may be toxic. For example, in the production of Humulin, the insulin that is made in bacteria, the objective is to remove one contaminating protein so that it is present at less than 0.1 parts per million in the final insulin preparation. Further processing may be required to obtain an active protein.

The hormone insulin is synthesized as a precursor called proinsulin. The proinsulin that is produced in bacteria must be modified by an enzyme to produce the active insulin hormone. Finally, the protein must be formulated in a manner that is suitable for the final application, e.g. preparation of insulin as an injectable solution. There are now many examples of proteins produced in this way, especially proteins that are used in the treatment of human diseases. Below are examples of some of these protein therapies that are already on the market:

- Treatment for diabetes (the first therapy derived from genetic engineering);

- Growth hormone - treatment for dwarfism;

- Interferon - for treatment of some cancers e.g. Kaposi's sarcoma and, multiple sclerosis;

- Erythropoietin - to treat anemia by stimulating bone marrow cells to produce red blood cells;

- Tissue plasminogen activator - to dissolve blood clots in victims of heart attacks and, more recently, in stroke patients;

- Hepatitis B vaccine - to vaccinate against this viral disease.

Microbes serve as ideal factories for producing pharmaceutical proteins for the following reasons: It is either difficult or impossible to purify these proteins from a natural source. For example, human growth hormone used to be produced from pituitary glands taken from human cadavers. It takes many donors to produce sufficient hormone to treat a single patient. Because of such supply limitations, the treatment was not widely available.

The proteins purified from microbes are much safer for the patients. When proteins are purified from donors, there is a significant risk that pathogens will not be removed and will be passed from donor to patient. An example of this is the treatment of haemophiliacs. In the past, clotting factors used to treat these patients were purified from donated blood. Before the virus responsible for AIDS was identified, it was impossible to test donated blood for this virus, HIV. Many (perhaps most) haemophiliacs who used clotting factors became infected with HIV because these were contaminated with the virus that causes AIDS. There are many other pathogens that can be transmitted in this way. Clotting factors and growth hormones purified from microbes are very unlikely to be contaminated with human pathogens, and this is one major advantage.

LITERATURE CITED

[41]Alex Albert (01/27/2009) – onging research at University of Minnesota.

[42]American Academy of Microbiology. *The Scientific Future of DNA for Immunization.* 1996. [ONLINE]: http://www.as-musa.org/acasrc/Colloquia/dnareprt.pdf [4/27/03 last day accessed].

[43]Ammon HPT, testa B (1996). In: Antidiabetic agents: recent advances in Their Molecular and Clinical Pharmacology – Academic Press.

[44]Court, Dr J. - Modern Living with Diabetes, Diabetes Australia, Melbourne, 1990.

[45]Chung S-M, SJ Advani, JD Bradely, Y Kataoka, K Vashitha, SY Yan, JM Markert, GY Gillespie, RJ Whitely, B Roizman and RR Weicheselbaum (2002). The use of a genetically engineered herpes virus (R7020) with ionizing radiation for experimental hepatoma. Nature Jan. 2002 vol. 9, No. 1, p. 75–80.

[46]Daniell H., Streatfield S. J., Wycoff K. (2001). Medical molecular farming production of antibodies, biopharmaceuticals and edible vaccines in plants. Trends Plant sci. 2001; 6: 219–26.

[47]David Templeton, Wed. April, 2007. Killing cancer from the inside out UPMC clinical trial uses genetically engineered virus in fight against liver cancer.

[48]Del Zoppo GT, Poeck K, Pessin MS, et al., (1992).

[49]Durham R. (1989), in: Human Physiology p. 337).

[50]Encke, J., Jasper zu Putlitz, and Jack R. Wands. 1999. DNA Vaccines. Intervirology 1999;42:117–124.

[51]Fauci Anthony S., MD; Giuseppe Pantaleo, MD; Sharilyn Stanley, MD; and Drew Weissman, MD, PhD (1996). Immunopathogenic Mechanisms of HIV Infection: annals.org [HTML]: **1 April 1996 | Volume 124 Issue 7 | Pages 654-663.**

[52]Flint, S.J. *et al.* 2000. Principles of Virology: Molecular Biology, Pathogenesis, and Control. ASM Press, Washington, D.C., 804p.

[53]Fischer R., Emans N. (2000). Molecular farming of pharmaceutical proteins. Transgen Res. 2000; 9:279–99.

[54]Fackelmann, Kathy A. (1989). Cancer-fighting tobacco plants? Science News, April 15, 1989.

[15]Goldstein, D. A. and J. A. Thomas (2004). Biopharmaceuticals derived from genetically modified plants. QJM vol. 97 No. 11 ©Association of Physicians 2004.

[55]Gustavo H. Kijak, Viviana Simon, Peter Balfe, Jeroen Vanderhoeven, Sandra E. Pampuro, Carlos Zala, Claudia Ochoa, Pedro Cahn, Martin Markowitz, and Horacio Salomon (2002). Origin of Human Immunodeficiency Virus Type 1 Quasispecies Emerging after Antiretroviral Treatment Interruption in Patients with Therapeutic Failure. Journal of Virology, July 2002, p. 7000–7009, Vol. 76, No. 14 0022-538X/02/$04.00+0. DOI: 10.1128/JVI.76.14.7000-7009.2002

[56]Guynup Sharon (2000). Seeds of new medicine-Genome News Network (GNN), July 2000. Copyright 2000–2004 J. Craig Venture Institue.

[57]Hellen Phillips, San Francisco (News Science (April, 2004): Alzheimer's gene therapy trial shows early promise. Extract from News science: http://www.newscientist.com/article/dn4930-alzheimers-gene-therapy-trial-shows-early-promise.html

[58]Hsueh, JL, Q Xingfang, Z Jiemin, D Yifan, H Yiping,..... - Human Gene Therapy, 1992 - liebertonline.com. HUMAN GENE THEKAPY 3:543-552 Ü992) Mary Ami Licbert. Inc.T Publishers Clinical Protocol Clinical Protocol of Human Gene Transfer for Hemophilia B.

[59]Jelaska Sibila, Sježana Mihaljevic, and Nataša Bauer (2004). Production of Biopharmaceuticals, antibodies and Edible Vaccines in transgenic Plants. *Current Studies of Biotechnology* – Volume IV. – Immuno-Modulatory Drugs http://www.hugi.hr/Current%20Studies%20of%20 Biotechnology%202005%20-%20Jelaska.pdf

[60]Kathryn senior (2003). Gene therapy with neprilysin reduces amyloid deposits in mice. The Lancet, Volume 361, Issue 9363, Pages 1107– 1107.

[61]Kiernan V. Yes, we have vaccinating bananas. *New Scientist* 1996 Sep 21: 6.

[62]Koprowski, H, and D.B. Weiner. 1998. DNA Vaccination/ Genetic Vaccination. Spriner-Verlag, Heidelberg, 198p.

[63]Kristina Collins (2007). Herpes virus to kill cancer cells? The CANCER BLOG July 12, 2007.

[64]Kumar Sunil GB, Ganapathi TR, Bapat VA (2004) Edible vaccines: current status and future prospects. Physiol Mol Biol Plants 10:37–47.

[65]Kumar Sunil GB; T. R. Ganapathi, C. J. Revathi, L. Srinivas and V. A. Bapat (October 2005). "Expression of hepatitis B surface antigen in transgenic banana plants". *Planta* 222: 484–493. doi:10.1007/s00425-005-1556-y.

[66]Maramorosch Karl, Fredrick A., Murphy Aaron J (1998). Advances in Virus Research. Science – 455 pages.

[67]Mason HS, Lam DMK, Arntzen CJ (1992). Expression of hepatitis B surface antigen in transgenic plants. Proc Nat Acad
Sci USA 89:11745–11749.

[68]maxen - 10 nov 08, 09:08
Crucell Vaccine Stops AIDS in Monkeys,
Harvard Scientist Says: By Simeon Bennett.
http://www.bloomberg.com/apps/
news?pid=20601202&sid=algk3S_Ei484

[69]Montefiori D, Sattentau Q, Flores J, Esparza J, Mascola J.
Antibody-based HIV-1 vaccines: recent developments and
future directions. *PLoS Med.* 2007;4:e348.

[70]Palmer TD, AR Thompson and AD Miller (1989). Production of
human factor IX in animals by genetically modified skin
fibroblasts: potential therapy for hemophilia B. Blood,
Volume 73, Issue 2, pp. 438–445, 02/01/1989.

[71]Peeters M., G.M. Shaw, P.M. Sharp & B.H. Hahn: Origin. of
HIV-1 in the chimpanzee Pan troglodytes troglodytes.
Nature 397, 6718 436–41 (1999).

[72]Prakash C. S. (1996). Edible vaccines and Antibody Producing
plants. Biotechnology Dev. Minitor 27:11–13.

[73]Raz, Eyal. 1999. Gene Vaccination: THEORY AND PRACTICE.
Springer-Verlag, Heidelburg, 169p.

[74]ScienceDaily (Oct. 27, 2007): Safer Gene Therapy? Hope for
Parkinson's Alzheimer's, and MS. Extract-from: http://
www.sciencedaily.com/releases/2007/10/071026095210.
htm

[75]Schirmbeck, R. and Jorg Reimann. April 2001. Revealing the
Potential of DNA-based Vaccination: Lessons Learned
from the Hepatitis B Virus Surface Antigen. Biol. Chem.,
382:543–552.

[76]Smith MF, Doyle JW. Use of tissue plasminogen activator to revive blebs following intraocular surgery. Br J Opthamalmol 2000; 84: 983–6.

[34]Snow, A. A., D. A. Andow, P. Gepts, E. M. Hallerman, A. Power, J. M. Tiedje, and L. L. Wolfenbarger. (2005). Genetically engineered organisms and the environment: Current status and recommendations. Ecological Applications 15:377-404.

[77]Travis John, Jan 30, 1999. Tobacco plants enlisted in war on cancer (TMV used to make vaccines) (Brief Article).

[78]University of Florida Health Science Center (2000, February): UF Dental Researcher Develops Genetically Altered Bacteria Strain That May Fight Cavities For A Lifetime. *ScienceDaily*. Retrieved May 21, 2009, from http://www.sciencedaily.com /releases/2000/02/000208135134.htm

[79]University of Massachusetts Amherst (2008, March 31). Salmonella Bacteria Turned Into Cancer Fighting Robots. *ScienceDaily*. Retrieved May 21, 2009, from http://www.sciencedaily.com /releases/2008/02/080229171124.htm

[8]Velander, H. William, Henryk Lubon & William N. drohan (1997). Transgenic livestock as drug factories. Scientific American January 1997, pages 70–74.

[80]Wallace Neal, (Oct. 2008). AgResearch health focus in GM work – web site.

[81]WASHINGTON, Nov 9 2008 (Reuters): Souped-up immune cells catch even disguised HIV: http://www.reuters.com/article/marketsNews/idINN0940487520081109?rpc=44

[82]Yale University (1999, December 13). Scientists Use Salmonella To Fight Cancer In First Human Trials. *ScienceDaily*.

Retrieved May 21, 2009, from http://www.sciencedaily. com /releases/1999/12/991213051204.htm

[83]Yale University (2001, January 11). Salmonella In Combination With Radiation Effective Against Cancer Tumors, Study By Yale Researchers Finds. *ScienceDaily*. Retrieved May 21, 2009, from http://www.sciencedaily.com / releases/2001/01/010111074337.htm

[84]http://en.wikipedia.org/wiki/HIV:-2009   [85]http://www.sciencemag.org/cgi/content/abstract/11   26531:- 2009.

CHAPTER IV

Transgenic Animals

The technologies used to produce GM animals (also known as transgenic animals) are well defined: Ref.: [87, 88, 91, 103, & 8]. The first transgenic mammal to be cloned [Dolly the sheep], was produced in 1996, following successful cloning from an adult cell – making her a genetic replica of a six-year-old ewe.

Methods designed to obtain transgenic animals such as mice, sheep, cows, and fish are fairly straight forward and involve certain principles. The principles include the barriers that the DNA must cross; whether or not the DNA of interest integrates into a chromosome; and using selectable markers to identify transformed cells. In producing transgenic animals, transformation of individual cells that are growing in tissue culture is first considered. Many different types of cells can be isolated from animals and grown in sterile culture, normally in a liquid medium contained in Petri dishes. These cell lines in culture are widely used for many different experiments that are set up for investigating the cellular and molecular biology of animal cells. There are a number of physical methods that can be used to introduce DNA into animal cells growing in culture. These methods include:

1. **Electroporation**
   Electroporation of animal cells is very similar to the procedure used for plant cells. The only significant difference is that animal cells do not contain a rigid cell wall, and so preparation of protoplasts (removal of the cell wall) is not required. Animal cells and DNA are mixed together and then subjected to an electric shock. It is believed that the electric shock produces small temporary pores in the membrane that allow DNA to enter the cell. At least some of the DNA can travel to the nucleus.

2. **Liposome - mediated uptake of DNA**
   Liposomes are artificial membranes that can be formed in a test tube. Using this method, DNA is mixed with the liposome preparation under the appropriate conditions.
   This results in the encapsulation of DNA into synthetic lipid membranes. When this membrane fuses with the cell plasma membrane, DNA is released into the cell and somehow ends up in the nucleus.

3. **Calcium phosphate precipitation of DNA**
   Calcium phosphate precipitates of DNA form when DNA is mixed with calcium chloride. When the precipitates are added to animal cells growing in culture, the precipitated DNA can be taken up by the cells, transferred to the nucleus, and expressed. The desired gene also known as transgenic gene is attached to a tissue specific promoter/regulatory region that drives transcription. The transgene can be expressed in many tissues of the transgenic animal. For example, a promoter containing the amylase gene confines expression to the pancreas. While an insulin promoter would restrict expression of transgene to islet beta cells.

For each of these methods, the DNA that is used for transformation is typically a plasmid that contains the gene of interest that you wish to transfer into these cells. The DNA comprising the gene of interest is frequently linked to a selectable marker gene which will allow the transformed cells to grow in the presence of an inhibitor such as an antibiotic. Each of these methods thus allows delivery of DNA to the nucleus of the recipient host animal cells. If stably transformed cells are needed, the cells are subjected to selection (use of a selectable marker) so that only transformed cells will grow. In this way we are able to select the transformed cells carrying the gene of interest.

It is also possible to use biological vectors to carry specific genes into animal cells. These vectors are derivatives of viruses that infect animals. These viral vectors can be divided into two groups:

1. Free replicating viruses that multiply within the cell, but do not integrate into the genome of the host.

2. Integrating viruses, including a group known as retroviruses (HIV and several others).

It is a common observation, that the free replicating viruses cannot produce stable transgenic cells, notably because the genetic material that is introduced does not become integrated into the host genome. However, this is still very useful for transient, not permanent, introduction of a gene. This idea holds the promise of treating some human diseases that are caused by genetic defects or modifications.

The integrating viruses such as HIV enter the cell, copy their RNA genome into DNA, and then this DNA integrates into the chromosome of the host. The genes carried on the virus can then be expressed, and also stably inherited after cell division. The integration of these retroviruses into the host genome is one reason why these viruses are so dangerous. Unless all the cells that carry the virus are removed, the host may still be attacked when the virus becomes re-activated.

It has been established that with either type of virus vector, not all of the cells in the host animal are transformed. Therefore, it is likely that not all the progeny of the infected animal will likely not be transformed, because some of the cells that give rise to gametes, eggs or sperm, may have not been transformed. Successful heritable transformation requires that the germ line cells are transformed. Unlike plants, it is extremely difficult to re-generate a fully differentiated animal from a single transformed cell growing in culture. This seemed like an impossible task but the arrival of Dolly, the first transgenic sheep, gave this technology hope; that it may be extremely difficult but not impossible.

There are a number of methods used to produce transgenic animals. Two of them are described: The first technique entails the injection of DNA directly into the nucleus of fertilized eggs (liposome mediated uptake). Mouse eggs are either fertilized in a Petri dish, or fertilized eggs are removed from an impregnated female mouse. Individual eggs are held stationary with a suction pipette. A few picoliters of DNA solution (about 1,000 molecules of DNA) are injected into one of the nuclei. The eggs are then implanted into a pseudo-pregnant foster mother. A total of 25-30 injected embryos are implanted. After 19-20 days gestation, transgenic baby mice are born. The transgenic baby mice (called founders) are identified by testing their genomic DNA (from tail region) for the DNA (gene) that was inserted. This can be done using a variety of methods such as PCR and DNA blot analysis (southern blot).

The heterozygous founders are then mated with non-transgenic mice. The resulting heterozygous siblings are again mated with each other to generate mice that are homozygous for the gene that was inserted. These animals constitute a transgenic line. This method of transformation is technically complex and efficiencies are not high. Reports show that, in mice and goats 90%

of the eggs survive injection of DNA, and about 20% of these are able to develop to full term upon re-implantation, while about 25% of these may be truly transgenic. In other mammals (e.g. sheep, pigs, cattle) efficiencies have been reported to be even lower than in mice. However, over time these techniques are becoming more routine and successful and efficiencies will increase.

The second method for producing transgenic animals involves transfer of transformed embryonic stem cells into embryos. Embryonic stem (ES) cells are the closest thing to totipotent cells that can be obtained from a mammal. The ES cells are isolated from young developing embryos. These cells are grown in culture to establish a cell line. Under appropriate conditions, the cells can develop into a variety of differentiated cell types. However, for transformation, ES cells are injected into a developing embryo where they give rise to a chimeric embryo. The animal that results from this chimera can be used to establish a line of transgenic mice. The ES cells can be transformed by any of the physical methods described above (electroporation, etc.). Transgenic ES cells are identified and propagated. The transformed ES cells are then injected into young blastocyst embryos. The resulting embryos will be chimeric mixtures of normal and transgenic cells, producing chimeric mice which are normally identified by their mottled appearance because of chimeras comprised of black and white cells. Chimeric mice are bred to identify offspring that have inherited the gene that was inserted.

Transgenic animals can be used to simulate certain diseases, and test new therapies. For example, we can have transgenic animals expressing the protein that controls development of Alzheimer's disease; or ventricular tachycardia. Ventricular tachycardia, an unusually fast heart rhythm, is the main cause of sudden death after a heart attack. Research suggests that transplanting genetically engineered cells into the heart of the patient may reduce the risk of a fatal condition which occurs after a heart attack.

The success of the gene therapy was supported by testing a variety of cells in mice that had been induced to have heart attacks. The study established that heart cells taken from 15-day-old embryos reduced the risk of ventricular tachycardia.

Transformation systems have been developed for many animals, including livestock (cows, pigs, and sheep) and chickens. One long term application of this transformation technology is to improve the quality of farm animals and reduce the cost of production in various ways. Attempts have been made to increase the growth rate of pigs by expressing a chimeric gene to increase growth hormone production in transgenic pigs. While pigs were produced that carried this gene, and also had increased muscle and reduced fat, these potentially beneficial traits were accompanied by a large number of deleterious abnormalities. It is possible that these negative effects were the result of this growth hormone being expressed in an uncontrolled manner in a wide range of tissues. It is therefore hoped that through gene manipulation, it would be possible to confer greater control over the tissues that express the growth hormone, which might lead to elimination of these problems.

Examples of traits that have been suggested as targets for biotechnology in animal production (e.g. sheep) include increasing the production of cysteine by expressing bacterial genes for two enzymes that convert serine to cysteine. Cysteine is a limiting factor in the production of wool in sheep. It has also been used to develop sheep expressing insecticidal proteins (perhaps Bt toxins) in sweat glands to reduce or eliminate the need for insecticidal sheep dips. This transformation technology has also been used in an attempt to manipulate disease resistance in animals by various methods. In cows, it has been used in altering the composition of milk. Targets for manipulation include increasing the accumulation of casein proteins (which are important for butter and cheese), reducing the level of whey proteins (which are of low value), and reducing the level of lactose (to which many people are intolerant).

GENETICALLY MODIFIED CHICKEN

Reports show that scientists have created the world's first breed of the so-called "new" chickens, genetically modified to lay eggs capable of producing drugs that fight cancer and other life-threatening diseases. Researchers at the Roslin Institue near Edinburg (where the first cloned sheep was created), have bred a large flock of the 'new' birds. The "new" egg-laying chicken, have each had human genes added to their DNA to enable them to produce complex medicinal proteins. The proteins with therapeutic properties are secreted into the whites of the birds' eggs, from which they can easily be extracted to produce drugs. The innovation by the Roslin scientists is reported to be a novelty; the birds that have been created "breed true", meaning the added human genes are passed on from generation to generation. Certainly, this opens the way for the creation of industrial-scale flocks and offers a potentially unlimited cheap source of medicinal proteins. [8, 110]

It is also reported that one of these chicken lines produces human interferon of a kind closely resembling a drug widely used to treat multiple sclerosis. Another line could be useful in treating skin cancer, by producing miR24, an antibody that could also potentially treat arthritis. However, this technology is still in its infancy, and it may take several years before the drugs made from the protein of eggs from these genetically-modified chickens can be safely used, if clinical trials show favorable outcomes. [8, 110]

GENETICALLY MODIFIED COWS

One area of animal biotechnology that has attracted a great deal of interest, and shows promise, is the use of animals to produce therapeutic proteins to treat human diseases. The strategy that is being used by many companies is to express high value proteins in the milk of large animals such as pigs, sheep or cows. Reports show that in the near future, dairy farmers may become suppliers of GM milk carrying selected therapeutic proteins; this is achievable through a biotech process known as biopharming. [8, 89, 94, 110]

It is widely acknowledged that milking is a well established procedure, and it should be feasible then to recover and purify the proteins from the milk in large quantities. In order to express a new protein in the milk of a transgenic sheep or cow, it requires the open reading frame for that protein to be used to construct a chimeric gene with the following components: A promoter to give a high level of expression in the mammary gland of the sheep is linked to the open reading frame of the target protein, and a transcription terminator. This leads to the production of transgenic animals that contain this gene. Females are then evaluated to see how much of the new protein is being made in the milk. Those females with the highest expression are then selected for further analysis, breeding to develop a herd of producers, and production scale-up. Reports show that Scientists at AgResearch have successfully bred cows that produced human myelin protein in their milk. Other examples of proteins that are being produced in this manner include:

- Various clotting factors to treat hemophiliacs;

- Tissue plasminogen activator to dissolve blood clots in heart attack and stroke victims;

- Alpha-1-antitrypsin to treat patients with emphysema.

At present, therapeutic proteins are mainly made in bio-reactors or vats of bacteria or other cells that have been genetically modified. However, extracting the relevant proteins is expensive and difficult. There are a number of advantages of using this approach of production in mammalian milk over microbe production systems. First, it may be possible to reduce production costs. However, the more important reason is that many proteins in eukaryotes are modified in ways that cannot be reproduced in microbes. Some of these biotechnological modifications include addition of complex sugars to specific amino acids, and cutting the amino acid chain into smaller peptides. [8, 89, 94]

Modifications like these are in many cases absolutely critical for the functioning of the protein. They can only be achieved when the protein is expressed in the mammary gland, but not in bacteria. Therefore, animal expression systems may have the advantage of producing an active functional protein, compared to what can be made in bacteria. A number of GM animals and plants are now being used as drug factories. One other possible future use of transgenic animals is to create herds of cattle or sheep that are genetically resistant to developing transmissible spongiform encephalitis, such as scrapie in sheep and "mad cow disease" in cattle. [8, 89, 94, 110]

GENETICALLY MODIFIED FISH

A successful example of development of transgenic animals comes from inserting a gene to over-express growth hormone in salmon. This resulted in a large increase in weight gain in the transgenic salmon. A lot of salmon are raised in aquaculture systems nowadays and the rate and efficiency with which the fish convert the supplied feed into marketable salmon fillets is very important to the financial success of these fish farming operations. [34, 102]

Reports show that fast-growing salmon that will increase production of farm-raised fish are currently being considered for commercial release. Research has shown that expression of introduced growth hormone genes, usually from the same species of fish, results in several-fold faster growth rates in salmon, tilapia, and mud loach. Reportedly, the transgenic fish have not yet been approved for consumption. The U.S. supermarkets and other consumer countries of GM foods currently sell no meat from genetically-engineered animals. A Boston-area company called Aqua Bounty Technologies, however, hopes to win approval for its faster-growing salmon and make the new fish available by 2011. [34, 87, 102, 105]

TRANSGENIC MOSQUITOES IMPAIRED IN TRASMISSION OF DISEASE PATHOGENS

Research on transgenic invertebrates has focused on control methods for insect pests and/or vector management for enhanced food security and public health applications, respectively. The production of vector insects that are incapable of transmitting disease presents a classical example of this. *Plasmodium* organisms are the intracellular protozoan parasites that cause malaria, and are vectored by female species of Anopheles mosquitoes. The *Plasmodium* depends on Anopheles mosquito to find human hosts. [95, 98, 99, 101, 106, 111]

Anopheles gambiae, a species that is indigenous to Africa, was the first mosquito species for which genetic transformation was reported [100]. Other Anopheline species have been genetically modified in an effort to eradicate malaria. However to date, there has been no field trial on the use of genetically modified mosquitoes to control malaria. The implied ethical, social and behavioral effects of such releases are currently under debate. The Anopheline that have been genetically modified include: *An. stephensi,* the major malaria vector on the Indian Subcontinent and *An. albimanus,* indigenous to South America. [104, 107] In addition, mosquito vectors of other diseases, for example *Aedes aegypti* (the vector for yellow and dengue fever viruses) and *Culex quinquesfasciatus,* (vector for West Nile virus and bancroftian filiariasis) are targets of genetic manipulation. With regard to control of malarial disease, recombinant gene technology is gaining popularity, in view of the growing concern caused by emergence and spread of drug resistance in Plasmodium parasites, insecticide resistance in mosquito vectors, and the absence of a vaccine. Therefore, gene manipulation of the mosquito vector, offers a good option as a long term prospect to eradicating malaria. This can be accomplished through interruption of the cycle of transmission, that is to say, by rendering mosquito vector species incapable of harboring the malaria parasite, thus making the vector incapable of spreading malaria. [86, 88, 95, 96, 97, 99, 106, 107, 108]

The genetic approaches that have been used are aimed at the ability of the mosquito to vector the disease. When a mosquito ingests a blood meal from a *Plasmodium*-infected human host, male and female gametocytes transform into gametes that mate, giving rise to zygotes. This takes place within the lumen of the mosquito's mid-gut. The zygote then transforms into an invasive form, the ookinete. Mature ookinetes attach to the surface of the mid-gut epithelium, migrate through it and anchor onto the basement membrane, and develop into oocytes. [98, 99, 101, 106, 107, 111]

Compelling evidence suggests that the progression of *Plasmodium* development in the mosquito critically depends on recognition of the epithelium of the mid-gut and salivary glands of the mosquito. Foreign genes can now be introduced into the germ line of the mosquitoes, and these transgenes can be expressed in a specific tissue, for example, the mid-gut or salivary gland epithelium, using promoters such as carboxypeptidase (CP). This can result in transgenic mosquitoes that express anti-parasite genes in their mid gut epithelium, thus rendering them inefficient vectors for the disease. Studies show that the CP promoter is strongly activated by a blood meal, and that the CP signal sequence drives secretion of the CP protein into the mid-gut lumen, where the initial stages of *Plasmodium* development takes place. The foreign gene that inhibits *Plasmodium* development for example, the pfpk7 gene which impairs schizogony and sporogony in the human malaria parasite *P. falciparum;* or CTRP (circumsporozoite protein and thrombospondin – related adhesive protein [TRAP.], is linked to the CP promoter, inserted into a transformation vector, and transformed into the germ line of the mosquito. Research has shown that, disruption of the TRAP gene results in severe reduction of malaria sporozoite motility and infectivity. [97, 98, 99, 101, 102, 104, 106 111]

The regulation of gene expression is controlled by regulatory proteins. The use of peptides for the genetic modification of mosquito vectorial capacity has been widely applied to shut down undesirable genes in a bid to control disease. For example, the mosquito can be engineered to express a salivary gland and midgut peptide, SMI derived from a carboxypeptidase promoter. The SMI binds specifically to the luminal side of the mid-gut epithelium and to the distal lobe of the salivary glands. Research has established that SMI significantly inhibits *Plasmodium* invasion of the two organs. Therefore, binding of the peptide to these epithelial surfaces interferes with parasite invasion. Previously, another peptide, the bee venom, phospholipase A2 (PLA2), has been shown to strongly interfere with *Plasmodium* invasion of the mid-gut. The genetically modified mosquitoes express the peptide from a carboxyptidase promoter. [98, 99, 101, 102, 104, 105, 106, 111]

Another genetic approach to control malaria is based on an endosymbiont of mosquitoes known as *Wolbachia* (Richettsiales: Rickettsiaceae). Wolbachia symbionts have been identified in many mosquito species, but no *Wolbachia* infections have been identified in any species of *Anopheles*. [97, 107, 108] The absence of *Wolbachia* in natural populations of *Anopheles* offers an advantage because preexisting natural *Wolbachia* infections are bound to interfere with, or interact with and alter the behavior of introduced infections. [90, 92, 95, 99, 107, 108, 110]

Studies have established that these endosymbionts (*Wolbachia*) are normally maternally inherited, and are generally associated with cytoplasmic incompatibility (CI), that is to say, they cause reduced egg hatch in matings between uninfected females and infected males. Some researchers suggest that *Wolbachia* symbionts impart certain changes to the affected male mosquitoes that favors their selective partnering up, and subsequent production of offspring with infected female mosquitoes. [105, 92]

In an attempt at gene manipulation, researchers have capitalized on this relationship between the mosquito and *Wolbachia*, by inserting a transgene (to confer plasmodial resistance) into the *Wolbachia* genome. *Wolbachia* can drive maternally inherited transgenes into mosquito populations and over time, the natural mosquito population would be replaced with one that is9more resistant to *Plasmodium* parasite transmission. [107] The *Wolbachia* symbiont can be introduced through embryonic microinjection; or injection of the symbionts into the adult insects. [92, 95, 99, 106, 107, 108, 110]

FUTURE PROSPECTS ON GENETICALLY MODIFIED ANIMALS

The realignment of our bio-tech reality has hit a new milestone, courtesy of a group of Japanese scientists. Reports show that the researchers have managed to create clones from the bodies of mice which have been frozen for 16 years. The Japanese research was undertaken at Kobe's Centre for Developmental Biology and is reported in the Proceedings of the National Academy of sciences (PNAS). The cloning was reportedly accomplished using barely live donor cells: the donor genetic material is reported to have come from animals stored at -20ºC for 16 years. Previous studies have shown that DNA or genetic material is un-affected by such harsh temperatures. [93]

According to the report, a normal cloning procedure entails the following: A cell taken from the animal to be cloned is fused with a donor egg, from which most of the genetic material has been removed. The whole idea is to get the genetic material of the cell of interest to be expressed (in favor of that of the donor egg) in the resultant embryo that will subsequently develop into an exact replica or clone of the original animal. In order to increase the chances of implantation of the cloned embryo into surrogate mothers, the Japanese researchers are reported to have used a modified technique of the above method; whereby after making the embryos using the direct injection technique, they obtained embryonic stem cells from those embryos and fused these with donor eggs thus creating cloned embryos with more vigor. The study was a success story according to the news brief; the mothers gave birth to healthy mouse pups. [93]

According to the scientists in Kobe, Japan, it is hoped their technique will raise the possibility of recreating extinct creatures, such as mammoth, from their frozen remains. Many of the successful clones that followed Dolly the sheep, have been created by a method where the nucleus of a cell has been removed, placed in an empty egg and kick-started into replicating using chemicals or electricity. Some of the major challenges faced by the researchers in applying this new technology include finding a suitable species to provide recipient eggs as well as surrogate mothers. [93]

LITERATURE CITED

[86]Anil, K. Ghosh, Paulo E. M. Ribolla, and Marcelo Jacobs-Lorena (2001). Targeting plasmodium ligands on mosquito salivary glands and mid-gut with aphage display peptide library. Proc. Natl. Acad. Sc. USA. 2001 November 6; 98 (23): 13278–13281.

[87]Devlin, R. H., T. Y. Yesaki, C. A. Biagi, E. M. Donaldson, E. M. P. S. Swanson, and W. K. Chan (1994). Extra-ordinary salmon growth. Nature 371: 209–210.

[88]Ghosh, A., Edwards, M. J. and Jacobs Lorena, M. (2000). The Journey of malaria in the mosquito: hopes for new century, Parasitol. Today 16, 196–201.

[89]Henninghausen, L. Transgenic factor VIII: the milky way and beyond. *Nature Biotechnology.* 1998;15:945–946.

[90]Hoffman, A. A., & Turelli, M. (1997). Cytoplasmic incompatibility in insects, in Influential Passengers (S. L. O' Neill, Ed.), pp. 42–80. Oxford University Press, Oxford.

[91] Jaerisch Rudolf (1988). In: Science, Vol. 240, pages 1468–1474: June 10, 1988.

[92]James, A. A. (2005). Gene drive systems in mosquitoes: rules of the road. Trends in Parasitology 21:64–67.

[93]Joe Palca (NPR.org, 2008): Scientists develop technique to clone Frozen mice. Research news: accessed on 17th Nov. 2008. GM food for thought Web site.

[94]Jonathan Leake (2007) – GM food for thought Web site.

[95]Junitsu Ito, Anil Ghosh, Luciano A. Moreira, Ernst A. Wimmer and Marcelo Jacobs-Lorena (2002). Transgenic anopheline mosquitoes impaired in transmission of a malaria parasite.

Nature [417], 452–455 (23 May 2002) | doi:10.1038/417452a; Received 28 December 2001; Accepted 11 March 2002.

[96]Kantoff, & HG Coon (1987). Stable integration and expression of a bacterial gene in the mosquito Anopheles gambiae. Science, Vol 237, Issue 4816, 779–781.

[97]Kittayapong, P., Baisely, K. J. Baimai, V., & O' Neil, S. L. (2000). Distribution and diversity of *Wolbachia* infections in Southeast Asian Mosquitoes (Diptera: Culicidae). J. Med. Entomol. 37:340–345.

[98]Kokoza, V. et al., (2000). Engineering blood meal activated systemic immunity in the yellow fever mosquito, A. aegypti. Proc. Natl. Acad. Sc. USA 97, 91444-9149.

[99]Marcelo Jacobs-Lorena (2003). Interrupting malaria transmission by genetic manipulation of anopheline mosquitoes. J. Vect. Borne Dis. 40, September & December 2003, pp 73–77.

[100]Miller, LH Sakai, RK, P Romans, RW Gwadz, P Kantoff, & HG Coon (1987). Stable integration and expression of a bacterial gene in the mosquito Anopheles gambiae. Science, Vol 237, Issue 4816, 779–781.

[101]Moreira, L. A. et al., (2000). Robust gut-specific gene expression in transgenic *A. aegypti* mosquitoes. Proc. Natl. Acad. Sci. USA 97, 10895–10898.

[102]Nam, Y. K. Noh, Y. S. Cho, H. J. Chow, K. N. Cho, C. G. Kim, and D. S. kim (2001). Dramatically accelerated growth and extra-ordinary gigantism of transgenic much loach *Misgurnus mizolepis*. Transgenic research 10: 353-362.

[103]Paleyanda Rekha, Janet young, William Velander and William drohin (1991). In: Recombinant Technology in Hemostasis and

Thrombosis. Edited by L. W. Hoyer and W. N. Drohan. Plenum Press, 1991.

[104]Perera, O. P., R. A. Harrel, & A. M. Handler (2002). Germ-line transformation of the South.

American malaria vector, *Anopheles albiminus,* with a piggy-back/EGFP transposon vector is routine and highly efficient. Insect Molecular Biol. 11291–7.

[105]Rahman, M. A., and N. Maclean (1999). Growth performance of transgenic tilapia containing an exogenous piscine growth hormone gene. Aquaculture 173: 333–346.

[106]Rasgon, J. L., Styer, L. M., Scott, T. W. (2003). *Wolbachia*-induced mortality as a mechanism to modulate pathogen transmission by vector arthropods, J. Med. Entomol. 40:125–132.

[107]Rasgon, J. L. & Scott, T. W. (2004). An initial survey *Wolbachia* (Rickettsiales: Rickettsiaceae) infections in selected Carlifornia mosquitoes (Diptera: Culicidae). J. Med. Entomol. 41:255–257.

[108]Ricci, L. Cancrini, G., Gabrielli, S. D' Amelio, S., & Favi, G. (2002). Searching for *Wolbachia* (Rickettsiales: Rickettsiaceae) in mosquitoes (Diptera; Culicidae): Large polymerase chain reaction surveys & new identifications, J. Med. Entoml. 39:562–567.

[34]Snow, A. A., D. A. Andow, P. Gepts, E. M. Hallerman, A. Power, J. M. Tiedje, and L. L. Wolfenbarger. (2005). Genetically engineered organisms and the environment: Current status and recommendations. Ecological Applications 15:377-404.

[109]Trurelli, M. & Hoffman, A. A. (1999). Microbe-induced cytoplasmic incompatibility as a mechanism for intro-

ducing transgene into arthropod populations. Genetics 132:713–723.

[8]Velander, H. William, Henryk Lubon & William N. drohan (1997). Transgenic livestock as drug factories. Scientific American January 1997, pages 70–74.

[110]Wallace Neal, (Oct. 2008). AgResearch health focus in GM work – web site.

[111]Zieler H, Keister DB, Dvorak JA, Ribeiro JM, (2001). A snake venom phospholipase A(2) blocks malaria parasite development in the mosquito mid-gut by inhibiting ookinete association with the mid-gut surface. J. Exp. Biol. 2001; 204: 4157–67.

CHAPTER V

Genetically Modified Foods

What constitutes a GM food? Well, genetically modified food (GM food) differs from non-GM (conventional) food in that it contains or is produced from genetically modified organisms (GMOs). Reports show that GM foods were first put on the market in the early 1990s. The most common GM foods are derived from plants: particularly maize, soya bean, tomato, papaya, potato, rice, oilseed rape, and cotton seed. Soya beans are also processed and used as an ingredient in a wide variety of foods, including chocolate products, bread, biscuits, infant formula, and breakfast cereals. Processed maize products can be found in many food products such as animal feed, maize flour, bread, and beer. In addition oil derived from genetically modified corn, cotton, rapeseed, and soya bean is used in cooking.

GM SEEDS WITH IMPROVED NUTRITIONAL VALUE

Seeds are an important source of protein but usually they don't contain the ideal ratio of amino acids, which are the building blocks of proteins. For example 50% of corn-seed protein is Zein, which lacks the amino acid lysine. Legume and tuber proteins are low in the sulfur amino acids, methionine and cysteine. This problem can obviously be overcome by eating larger quantities or mixing various kinds of foods, which is costly. Other shortcomings of eating plant proteins; is that some of the proteins contain biologically active compounds such as trypsin inhibitors, and cyanogenic glucosides. These compounds are present in plant seeds, roots, tubers, and leaves. Some of these compounds are toxic if the plant produce is not properly processed. For example, cassava tuber contains cyanogenic glucosides which require a tedious process to remove or detoxify. [116, 117, 125]

Using biotechnology, the amino acid content of soya beans has been improved; and plants expressing lower amounts of toxic and anti-nutritional compounds have been developed. This technology has also been used in the production of corn and wheat with improved lysine content thus making for an imbalance of amino acids in human and animal nutrition. [125]

GENE MANIPULATION TO REDUCE BROWNING IN FRUITS & VEGETABLES

It is a common observation that most fruits and vegetables change color or turn brown when they are freshly cut and left to stand for long. This is a big problem for fruit and vegetable vendors as well as many restaurants. However, most restaurants control this natural browning phenomenon using food preservatives such as sulphur dioxide. Reports indicate that this chemical has been implicated in asthma-related deaths arising from eating food containing the preservative. [21, 124, 129]

Studies have also shown that the browning reaction in freshly cut plant tissue is due to the oxidation of phenolic compounds in the fruit; brought about by the release of the of the enzyme polyphenol oxidase (PPO) by the damaged plant tissues. This enzyme naturally occurs in the plant tissues and catalyses the conversion of monophenols (released from separate sub cellular compartments) to quinones, which then oxidise to form brown polyphenolic pigments. The gene for polyphenol oxidase has been switched off in experimental studies by genetic modification, blocking the browning phenomenon in several fruits and vegetables including potatoes, apples, beans, grapes, sugarcane and lettuce. [21, 124, 129]

OTHER GENETIC MODIFICATIONS THAT IMPROVE FOOD QUALITY

has shown that the balance of sugar and starch in potatoes, which affects the processing quality for snack food production (too much sugar produces a dark, poor tasting product), can now be modified on an experimental scale. Modern genetic techniques are being used to identify and manipulate the genes for biologically active components of food crops including natural toxicants such as potato glycoalkaloids; kidney bean lectin; antinutrients such as trypsin inhibitors; and allergens such as nut proteins. Such developments are at early stages but will certainly lead to the production of foods that lack these undesirable components. [114, 117, 125]

This new technology also has the added advantage of producing foods with greater benefit to man, such as wheat with increased fiber to reduce risk of colon cancer; and wheat with increased folic content to prevent spina bifida. Some "new" bruise-tolerant potatoes have been designed to have an increased starch content and absorb less oil. However, reports show that all these developments are at the experimental stage and will take a long time before they are approved for commercialization. [114, 117, 125]

"GOLDEN RICE" - Production of Vitamin A in Rice

The so called "Golden rice" is a variety of colored (yellow-orange) rice in which the genes for the production of beta-carotene have been inserted. Beta-carotene is a precursor of vitamin A; the human body can use it to form the vitamin. Researchers have inserted three genes (two from maize or daffodil plants and one from a common soil bacterium - *Erwinia)* into the kernel (the endosperm) of rice to allow the plant to produce beta carotene. [113, 127, 128, 131]

Studies have shown that the enzyme products of these genes lead to the formation of a substance known as lycopene, a process that does not naturally occur in the endosperm of rice. The lycopene produced is then converted into beta carotene and other provitamin A carotenoids (the building blocks of vitamin A) by other enzymes found in the rice grain. These findings are important because vitamin A (retinol) deficiency is the world's leading cause of blindness; it is estimated that it affects millions of children from developing countries, especially in Asia and Africa. Examples of natural sources of vitamin A include butter, fatty-fish liver oil such as Cod liver oil and sheep's liver. [113, 120, 127, 128, 131]

Provitamin A compounds such as beta carotene are found in dark green vegetables, fruits and tubers. The rice plant naturally makes beta-carotene and other carotenoids, which are present in the entire plant, except in the endosperm. The endosperm constitutes white rice that is commonly produced and eaten. Researchers suggest that the genetic manipulation that produces golden rice is simply designed to extend this natural production of beta-carotene into an additional part of the plant. However, eating golden rice may not necessarily alleviate vitamin A deficiency. This is due to the fact that beta-carotene is fat-soluble and its uptake by intestines depends upon fat or oil in the diet.

White rice itself does not provide the necessary fats and oils, and poor, malnourished people usually do not have ample supplies of fat-rich or oil-rich foods. Subsequently, if they were to eat golden rice without fats or oils, much of the beta carotene would pass undigested through the intestine. Moreover, the process of translocation of carotene and vitamin A from the intestines to the liver for storage also requires enzymes (which are proteins), in addition to fats. Proteins bind to the vitamin in the liver while enzymes transport it to the different body tissues where the vitamin is utilized. A person who suffers protein related malnutrition and lacks dietary fats and oils will have a disturbed vitamin A metabolism. [119, 120, 128, 127, 128]

Reports show that the process that led to the successful development of this "new" rice has been long (more than twenty years) and tedious, and that it may take several more years to advance the Golden rice product through the regulatory approval process. [127, 128]

GENETICALLY MODIFIED MILK & MILK PRODUCTS

One of the first uses of proteins produced in bacteria has been to increase milk production in cows. It has been known for many years that injection of bovine somatotropin (BST, a growth hormone) into lactating cows could produce a significant increase in milk production. However, it has been impossible to obtain sufficient BST to do this on a commercial scale. [115, 118]

Recombinant Bovine Growth Hormone (rBGH - also known as Bovine Somatotropin, or BST) is a genetically engineered copy of a naturally-occurring hormone produced by cows. It works by altering gene expression of glucose transporters in the cow's mammary gland, skeletal muscle and developmental fat. The gene facilitates the repartitioning of glucose to the mammary gland, which in turn produces more milk. Reports show that it is possible to produce this hormone in bacteria, and inject cows with a daily dose. BST produced in this way is now sold by at least two companies in the United States, Monsanto. However, there has been a great deal of controversy about the use of BST, the effect of this on cows, and the safety of the milk produced from treated animals. [115, 118, 121, 123]

Genetically Engineered Rennet

Rennet is a natural complex of enzymes produced in any mammalian stomach to digest the mother's milk, and is often used in the production of cheese. Rennet contains many enzymes, including a proteolytic enzyme (protease) that coagulates the milk, causing it to separate into solids (curds) and liquid (whey). The active enzyme in rennet is called chymosin or rennin, but there are also other important enzymes in it, e.g., pepsin or lipase. There are various sources of rennet. Natural calf rennet is extracted from the inner mucosa of the fourth stomach chamber of young calves. The stomachs are a by-product of veal production. If the rennet is extracted from older calves the rennet contains less or no chymosin but a high level of pepsin can only be used for special types of milks and cheeses. As each ruminant produces a special kind of rennet to digest the milk of its own mother, there are milk-specific rennets available, such as Kid goat rennet especially for goat's milk and lamb rennet for sheep's milk. Rennet or digestion enzymes from other animals, like swine-pepsin, are not used in cheese production. [130]

Because of the limited availability of proper stomachs for rennet production, cheese makers have looked for other ways to coagulate the milk. There are many sources of rennet enzymes, ranging from plants, fungi, and microbial sources, that can be used as a substitute for animal rennet. Cheeses produced from any of these varieties of rennet are suitable for vegetarian consumption. Many plants have coagulating properties. For example, an extracts of fig, nettles, thistles, mallow and other plant juices have been used to coagulate milk. Enzymes from thistle or cyanara are used in some traditional cheese production in the Mediterranean. [130]

Worldwide, there is no industrial production for vegetable rennet. Commercial so-called vegetable rennets usually contain rennet from the fungus *Mucor miehei*. The flavor and taste of cheeses produced with microbial rennets tend towards some bitterness, especially after longer maturation periods. [112] Because of such imperfections of microbial rennets, some producers sought further replacements of natural rennet. With the development of genetic engineering, it became possible to use calf genes to modify some bacteria, fungi or yeasts to make them produce chymosin. Chymosin produced by genetically modified organisms was the first artificially produced enzyme to be registered and allowed by the U.S. Food and Drug Administration. [19, 21, 130]

The gene for bovine chymosin has been transferred to industrial microorganisms – *Kluyveromyces lactis* (a yeast), *Aspergillus niger* var awamori (a fungus), and *Escherichia coli* K12 (a bacterium). These microorganisms are grown in fermenters to produce chymosin (rennet) on a commercial scale. This so called genetically modified rennet has replaced the conventional form obtained from slaughtered animals, and is now widely used in cheese production. [21, 121, 123] Today the most widely used genetically engineered rennet is produced by the fungus *Aspergillus niger.* [21, 121, 123]

Researchers have established that Cheese production with genetically engineered rennet is similar to production with natural calf rennet. Often a mixture of genetically engineered chymosin and natural pepsin is used to imitate the complexity of natural rennet and to get the same results in coagulation and in the development of flavour and taste. The so-called "GM rennets" are suitable for vegetarians if there was no animal based alimentation used during the production process. GMO-Microbial rennet is gaining popularity because it is less expensive than animal rennet. [7, 19, 130]

GENETICALLY MODIFIED TOMATOES

Ref.: [122, 126]

Many transgenic plants have been developed to improve the quality of the product. For example, genes have been inserted into tomato plants, which delay tomato ripening and softening. As a result, these tomatoes have a longer shelf life, and spoilage is reduced. These genetically altered tomatoes are referred to as Flav'r Sav tomatoes.

Understanding the process of fruit ripening will enable us know the genetic process used to modify tomato ripening, the penultimate stage in fruit development. After fertilization, the tomato fruit grows, first through cell division and then by cell expansion, until it reaches its maximum size. This takes about 40 to 50 days. The mature green fruit then undergoes a number of dramatic changes: increase in ethylene production, increased respiration, synthesis of the red pigment (lycopene), softening of the fruit, development of flavors, conversion of starches to sugars, etc. All these processes are regulated. Ripening is thus not just the random deterioration of biological processes in the fruit. Previous studies have shown that the action of certain genes is required for normal fruit ripening and when the genes are inactive, as a result of a mutation, ripening does not occur.

Specific biochemical processes are activated during fruit ripening. The first question to ask about fruit softening is what makes the fruit firm in the first place. Firmness of the fruit is a function of the properties of cell walls, the structural component that surrounds every plant cell. Cell walls are comprised of: cellulose fibers; pectins; hemicelluloses; proteins. The cross-linking components of the cell walls (pectins, Hemicelluloses, cellulose fibers and proteins) make the cellulose fibers more rigid.

During fruit ripening, various enzymes that degrade specific components of the cell wall are synthesized in the fruit. Among the enzymes that accumulate in the fruit are cellulases (to break down cellulose), and polygalacturonase (PG) and pectin methylesterase (PME), both of which are involved in breakdown of the pectin cross-linking molecules. These enzymes contribute to the softening of the fruit by reducing the rigidity of the cell wall structures. The expression of the genes encoding these enzymes is regulated by ethylene. These gene products, usually proteins, are then responsible for the specific changes that occur during the ripening process. When synthesis of ethylene is blocked using chemical inhibitors of the enzymes involved in ethylene synthesis, tomato fruits will not ripen. Apart from controlling the fruit ripening process, ethylene is also involved in other facets of plant growth, including seed germination and seedling growth, leaf abscission, petal senescence, and responses to environmental stress.

Ordinarily, ripe tomatoes cannot be transported without getting damaged. The ripening process can be halted by inhibiting ethylene biosynthesis. Two varieties of Flav'r Sav tomatoes are now on the market engineered for delayed ripening: One variety has an extra gene, a reverse copy of the gene responsible for an enzyme that breaks down cell walls, thus slowing down the softening process.

The other variety has a gene that controls the enzyme necessary for the production of ethylene, one factor that makes a tomato soft. This tomato variety has been obtained using antisense strategy, whereby a DNA sequence producing RNA complementary to the mRNA encoding an enzyme in ethylene biosynthesis is introduced in tomato plants.

The antisense RNA strategy is designed to accomplish the following: First, the plant must be engineered to produce an RNA molecule that has the complementary sequence of bases as the messenger RNA whose expression we wish to abolish. This RNA is known as antisense RNA. In order to manipulate a plant and be able to produce antisense RNA, transgenic plants have to be made containing a new chimeric gene which will make anti-sense RNA when it is transcribed.

In a normal gene, the promoter directs the enzyme RNA polymerase to transcribe mRNA, using the lower strand of the DNA as a template to make an RNA that has the same sequence as the upper strand of the DNA. In this case the base Uracil replaces Thymine. The transgene used to make antisense RNA makes a transcript of the same piece of DNA, but instead of using the bottom strand as template, it uses the upper strand and is transcribed in the opposite direction. The result is that the RNA transcribed form this "antisense RNA gene" has the same sequence as the lower strand.

The mRNA is inactivated by making it double; when it combines with complementary strand (the ribosomes cannot translate it). Because of the complementary sequences in the messenger RNA (also called sense RNA) and the antisense RNA, the two RNA molecules will anneal by hydrogen bonds forming between bases and form a double stranded RNA molecule. Obviously, when the mRNA is in this double stranded form, it cannot be translated into protein. Therefore, antisense RNA can block expression of a gene by preventing translation from ever occurring. This is the basis for using antisense RNA to inactivate the expression of specific genes, including ethylene biosynthesis in plants. Once these double stranded RNA molecules are formed, it appears they are rapidly degraded by some unknown mechanism within the plant cell. In normal fruit, ethylene production is described as being autocatalytic, where exposure to ethylene stimulates production of more ethylene. However, in transgenic fruit these mechanisms to control ethylene synthesis prevent the stimulation of ethylene synthesis by ethylene.

A second aspect of tomato fruit ripening that is a target for modification is the softening of the fruit. As mentioned earlier, the variety of Flav'r tomatoes has an extra gene, a reverse copy of the gene responsible for an enzyme that breaks down cell walls, thus slowing down the softening process. Research has shown that reducing the expression of polygalacturonase (PG) in tomato fruit would slow down the softening of the fruit. They used the antisense RNA approach to produce plants with reduced expression of PG in fruit. It was previously claimed that softening of the fruit was slowed in these fruit, allowing them to remain on the vine longer, with harvest later than the typical "mature green fruit" stage. The tomatoes could then be shipped and marketed before they turned to mush. However, reports show that reducing expression of PG did not have the expected effect on fruit softening.

Other researchers have tried essentially the same antisense RNA approach to produce plants with low expression of pectin methylesterase (PME) in the fruit, again using antisense RNA. As its name implies, PME is involved in metabolism of pectins in the cell wall. Pectins in mature green fruit are long polymers, and PME is expressed during fruit ripening to break these large polymers into shorter molecules. It is likely that PME is one of the first enzymes involved in the metabolism of pectins. It has been established that the pectins in these fruit with reduced PME activity remain large as the fruit ripen. They are not broken down into shorter pectins because there is no PME activity. As a result of this change in metabolism during ripening, the pectins remain as large polymers.

Previous research has established that Long polymers increase the viscosity of solutions. Juice extracted from low-PME fruit has a higher viscosity than juice from normal plants. And, the outcome of this modification is that using this more viscous juice as the starting material means less processing is required to produce tomato paste of the desired consistency. This has great potential for reducing the cost of processing these tomatoes, and perhaps improving the quality of their final product. Further research is in progress to see if tomato fruit with reduced expression of PME can be developed into a successful product.

While the fruit with reduced expression of PG were not successful in improving the quality of fresh tomatoes; it has been observed that reducing the expression of PG in fruit does increase the soluble solids content of tomatoes, likely by increasing the size of pectins. Tomato paste processed from these fruit is currently on the market in Britain. In addition, Flav'r Sav tomatoes are FDA approved, and they are on the market in the USA.

LITERATURE CITED

[112]Agboola Samson, Shaojiang Chen, and jian Zhao (2004). "Formation of bitter peptides during ripening of ovine milk cheese made with different coagulants" (in English, French). Lait (EDP Sciences) 84: 567–578. Doi:10,1051/ Lait:2004032 (inactive 2008-06-25). http://www.lelait-journal.org/index.php?option=article&access=standar d&Itemid=129&url=/articles/lait/abs/2004/05/L0420/ L0420.html. Retrieved on 2007-12-31.

[113]Alan Dove *Nature Biotechnology* 18, 135 (2000) doi:10.1038/72531: Golden rice.

[114]Bachem, CWP; Speckmann, GJ; van der Linde, PCG; Verheggen, FTM; Hunt, MD; Steffens, JC, et al. Antisense expression of polyphenol oxidase inhibits enzymatic browning in potato tubers. *Biotechnology.* 1994;12:1101–1105.

[115]Burton, J. H., G. K. MacLeod, B. W. McBride, J. L. Burton, K. Bateman, I. McMillan, and R. G. Eggert. 1990. Overall efficacy of chronically administered recombinant bovine Somatotropin to lactating dairy cows. J. Dairy Sci. 73:2157–2167.

[116]2 BBSRC Business, January, 1998,"Making crops make more starch" P.6–7.

[117]Chaomin Meng, Xüquing Chen, Rongqi Liang, Fenping Yang, Liquan Zhang, Xiaodong Zhang, Tianyou Chen, and S. S. M. Sun (2004). Expression of lysine-rich protein gene and analyzing lysine content in transgenic wheat. Chinese Science Bulletin, Vol. 49, Number 19/October 2004.

[118]Challacombe, D.N. and Wheeler, E.E., 'Safety of milk from cows treated with bovine somatotropin', The Lancet, September 17, 1994, Vol. 344, p.815.

[119]Erdman, J. et al (1993). Absorption and transport of Carotenoids, in Carotenoids in Human Health, edited by L. Canfield et al. New York: New York Academy of Sciences.

[120]Guerinot, M. (2000). The Green Revolution Strikes Gold. Science 287:241–243.

[19]Institute of Food Science and Technology (IFST, 2008). Genetic Modification and Food p. 1–44.

[121]Jousan, F. D., L. A. de Castro e Paula, J. Block, and P. J. Hansen (2007). Fertility of Lactating Dairy Cows Administered Recombinant Bovine Somatotropin During Heat Stress. J Dairy Sci, January 1, 2007; 90(1): 341–351.

[21]Leighton Jones, publication manager: BMJ. 1999 February 27; 318 (7183): 581–584. Copyright © 1999, British Medical Journal - Science, medicine, and the future - Genetically modified foods.

[122]Martineau, Belinda (2001). First Fruit: The Creation of the Flavr Savr Tomato and the Birth of Biotech Foods. McGraw-Hill, 269. ISBN 978-0071360562.

[123]Mepham, T.B., et al., 'Safety of milk from cows treated with bovine somatotropin', The Lancet, November 19, 1994, Vol. 334, pp. 1445–1446.

[124]Michelmore R. W. & Trevor Suslow, (11/15/97): ICEBERG LETTUCE – UC Vegetable Research and Information Center.

[125]Monsanto Corporation news brief, extracted from: www. laleva.org/eng/2005/11/terminator_comeback_in_monsanto_high_lysine_gm_corn.html - 30k

[126]Oeller, P. W., Min-wong, L., Taylor, L. P., Pike, D. A. & Theologis, A. (19910. Inhibition of tomato fruit senescence by anti-sense RNA. Science 254: 437–439.

[127]Paine et al., (2005). Nature Biotechnology 23, 482–487.

[128]Potrykus, I. "Golden Rice and Beyond." *Plant Physiology* 123 (March 2001): 1157-1161.Prakash, C.S. (1996), "Edible Vaccines and Antibody Producing Plants." *Biotechnology and Development Monitor*, No. 27, p. 10–13, June 1996.

[7]Smith, J. E. (2006). Public Perception of Biotechnology. Edited by Colin Ratledge and Bjorn Kristiansen. 3rd Ed. Cambridge, NY, Cambridge University Press, 2006. P. 3–33.

[129]Tim Thwaites, 21 January 1995: Wave goodbye to discolored fruit – Magazine issue 1961.

[130]Wikipedia ⁻ Rennet: **http://en.wikipedia.org/wiki/ Rennet**

[131]Ye, X. et al., (2000). Engineering the provitamin A (Beta-carotene) Biosynthetic pathway into (carotenoid-free) rice Endosperm: Science: 287:303–305.

CHAPTER VI

Transgenic Plants

Transgenic plants can be used for a variety of applications including improvement of product qualities, increasing yields, producing plants that are resistant to pests and diseases, and producing drugs. Recent biotechnological tools of direct gene transfer help engineer into plants new characters that are otherwise very difficult to transfer by breeding programs. Since 1983, when the first successful transformation of a plant cell with a foreign gene and the regeneration of a fertile plant from this cell was reported, numerous plant species have been transformed. These include important horticultural and field crops such as cucumber, eggplant, lettuce, melon, pea, maize, rice, oilseed rape, potato, tomato, strawberry, cotton, coffee, soya bean, rye, sugar beet, sunflower, and tobacco. [139, 141, 150, 153]

Today, transgenic plants resistant to viruses, bacteria, fungi, insects, and herbicides exist. Also available are transgenic plants which produce higher amounts of nutritious seed proteins, such as methionin-rich proteins or those engineered to produce biologically active compounds used in the pharmaceutical industry. Hybrid seeds can now be obtained through genetic engineering of nuclear male sterility while the ripening of fruit can be controlled and the color of flowers can be changed through the genetic engineering of antisense genes. [16, 36, 133, 139, 139, 150]

HERBICIDE RESISTANT PLANTS

Ref.: [135, 152]

The first transgenic plant was a herbicide resistant plant. Roundup, a non-selective herbicide also known as glyphosate, is known to kill weeds but may also kill crops.

In order for a herbicide to selectively kill weeds but not affect a crop, the crop plants must have some mechanism of preventing the herbicide from causing the damage that it inflicts on weeds. Which mechanisms do plants use to survive or tolerate a herbicide? To answer this question, let us consider what the herbicide must do in order to be effective.

Glyphosate is a very effective systemic broad spectrum herbicide. It is readily taken up through the leaf, and is transported throughout the plant. There is some concentration and accumulation of the herbicide in shoot meristems, and these tissues are among the most sensitive to glyphosate. Glyphosate inhibits the enzyme EPSP synthetase that is involved in making one group of amino acids. Unlike humans and many other animals, plants synthesize from scratch all the amino acids that are required for protein synthesis and other metabolic functions. Glyphosate therefore works by inhibiting amino acid biosynthesis and prevents plants from making aromatic amino acids. Obviously, if a plant cannot make amino acids that are required for protein synthesis, it will die. The way glyphosate works to inhibit plant growth indicates why it is such an effective and relatively safe herbicide. All plants (and microbes) possess both this enzyme and the pathway to synthesize aromatic amino acids and are therefore susceptible to glyphosate. Another important point to note is that glyphosate is not metabolized very quickly by plants, unlike when it is present in soil where it is immobilized and metabolized by soil microorganisms. The enzyme that is inhibited by glyphosate is not present in animals. Animals do not have a shikimate pathway but obtain aromatic amino acids in their diet.

In an effort to engineer a plant resistant to glyphosate, scientists isolated a bacterium that was resistant and cloned the resistance gene which turned out to be an altered EPSP synthatase gene (called AroA). They took the AroA gene from the resistant bacteria and inserted it into the plant leading to the production of the glyphosate resistant gene in the plant. This method was used to transform tobacco plants. The fields planted with the glyphosate resistant plants could be sprayed with glyphosate to kill all the weeds without killing the tobacco plants themselves.

A general observation is that these new biotechnology crops could offer many advantages, including the following: Herbicide resistant crops will allow farmers to spray their fields either before the crop is planted or when the crop is still young. This will result in the use of less herbicide. Reports show that in addition to decreased use of herbicide, planting herbicide resistant crops leads to less soil erosion. When herbicide sensitive crops are planted, fields are tilled, weeds allowed to grow, sprayed and then fields are left for the herbicide to dissipate (during which time rain can wash top soil away), before crops are planted. In the case of herbicide resistance, the advantage is that crops and weeds grow up together, are both sprayed and the dead weeds can act as mulch.

Potential dangers of engineering plants with herbicides include:

- Increased use of herbicide brings up human health issues. It has been established that humans do not contain the AroA gene, and are therefore not affected by the glyphosate.

- Most herbivores do not have the enzymes that synthesize amino acids and are therefore not harmed by ingesting the transgenic plants.

- Resident time of the herbicide: Glyphosate breaks down quickly, but other herbicides remain in plants for longer period.

- The general fear that gene transfer could create weeds that are resistant. Many agricultural crops grown world wide are hybrid and therefore don't outcross very well. This may not be a serious danger.

- Vertical control by industry: The farmer has to buy seeds with processed genes and herbicide from one source. The fear of creating monopoly by big companies.

TRANSGENIC PLANTS RESISTANT TO MICROBIAL PATHOGENS

Ref.: [133, 143, 149]

Microbial pathogens are estimated to cause a reduction in crop production of approximately 12% overall. This translates into an estimated annual loss of approximately $42 billion dollars worldwide, quite a major impact. Under some conditions these pathogens can cause epidemics that result in the total loss of a crop. In addition to reduced crop yields, the quality of the crop is also affected.

Reports show that the increasing use of monoculture systems in agriculture is also reducing genetic diversity and can contribute to the problem of susceptibility to diseases. An excellent example of this, cited in literature, occurred in the US in the 1970s. A disease called Southern Corn Leaf Blight was causing dramatic losses in the US corn production. The reason for the appearance of this disease lay in the use of a particular genetic background that was very widely used to produce hybrid corn seed. A strain of the pathogen appeared that was particularly damaging to corn hybrids that were developed in this background, which was used on a large fraction of the corn acreage. However, once the seed companies engaged a different method of producing hybrids that didn't rely on this disease susceptible genetic background, the problem went away.

The current methods used to control bacterial and fungal pathogens include:

1. Cultural practices: In a horticulture greenhouse operation, soil mixes are usually sterilized by high temperatures to eliminate pathogens that pose problems especially for germinating seeds.

2. Planting varieties with genetic resistance to specific pathogens. These genetic determinants may consist of a single gene or be controlled by many genes.

3. Using pesticides, especially fungicides.

All plants are exposed to, and are probably covered by, hundreds or thousands of different bacterial species. Some of these organisms are pathogens; they attack the plant causing diseases, whereas others do not cause any disease and are not pathogens. In order for a disease causing organism to be a pathogen, it must be able to penetrate the physical barriers of the plant such as the cuticle and cell wall. Once the pathogen has entered the plant, it may lyse (burst) plant cells and feed on the nutrients that are available from the plant. Plants have also developed various strategies and mechanisms to fight pathogens. These include:

1. Upon infection by a pathogen (bacterium, fungus or virus) plants respond by synthesizing a small number of new proteins. These are called pathogenesis-related (PR) proteins. Some PR proteins have antimicrobial activity, i.e. they inhibit the growth of these organisms. Examples of PR proteins include chitinases and glucanases. These are able to digest the cell walls of fungi; a fungus without a cell wall will not live for very long. Other PR proteins are able to make the membranes of some bacteria leaky. The PR proteins are synthesized both at the site of infection as well as systemically. A signal is transmitted to the rest of the plant which leads to systemic production of the proteins throughout the plant. At least part of the signaling is carried out by a compound known as salicylic acid. Regardless of how the plant sends this signal from the site of infection, the result is that plants are protected from subsequent infections by the same or different pathogens. This response is known as systemic acquired resistance (SAR).

2. Plants also respond to pathogens by inducing the synthesis of phytoalexins. These are antimicrobial compounds that are usually derived from what is known as "secondary metabolism."

3. Lastly, plants may restrict the spread of a disease by sacrificing a number of cells around the site of infection. This is referred to as the hypersensitive response or HR. Dead cells do not provide the necessary nutrients for pathogens to grow. Research has established that this process of cell death is highly controlled and includes production of hydrogen peroxide to kill the pathogen; and thickening of cell walls surrounding the pathogen to restrict its growth.

Biotechnology Strategies for Improvement of resistance to pathogens
Ref [138, 147, 151]

The following are the biotechnological strategies that have been developed:

1. 1). Constitutive expression of PR proteins: Since PR proteins have some antimicrobial properties, it was thought that expressing one of them all the time (constitutively) might be able to protect plants from pathogens. Transgenic plants have been produced that express many different PR proteins. In most cases, these individual PR proteins have little or no effect. In some cases it could be shown in the laboratory or greenhouse that these plants had some modest increase in resistance to one or two pathogens, but this was never enough to be effective in the field.

Constitutive expression of two or more PR proteins may be much more effective than single PR proteins. Such combinations may have a synergistic effect, that is to say, the combination is better than you would expect if you just added together the effects seen with both expressed independently.

Examples of the combinations that have been used are chitnase and glucanase, and a ribosome inactivating protein (RIP) with a cell wall hydrolase.

2. Expression of antimicrobial proteins and peptides: Many organisms other than plants produce proteins or small peptides with antimicrobial activity. These compounds either slow down growth of microbes or kill them. Because these are composed of a chain of amino acids, they are encoded by genes. A variety of genes expressing these antimicrobial peptides or proteins have been expressed in different transgenic plants. Among those that have been tested are defensins from plants, cecropins and attacins from insects, lysozyme from chicken or other sources. Lysozymes kill by breaking down the cell walls of the pathogen.

3. Increased production of hydrogen peroxide: The enzyme glucose oxidase (from Aspergillus) produces hydrogen peroxide as part of its normal activity. This gene has been expressed in potato with some clear effect on resistance to bacteria that cause soft rot disease. The hydrogen peroxide is able to restrict the growth of the pathogen and also activate the expression of some PR proteins.

4. Expression of resistance genes: Resistance genes are responsible for resistance to some pathogens. In general, it has been found that they are effective against only a small subset of pathogens rather than providing resistance to all pathogens. Over the last few years, a number of resistance genes have been cloned. In a small number of cases, these have been transferred between species and have some effect on resistance to a pathogen.

5. Expression of a "death gene": Using this approach, a chimeric gene has been constructed that consists of the promoter from a PR protein gene and an open reading frame encoding a ribonuclease (RNase) enzyme that destroys RNA. This gene is primed to be expressed in cells that are infected with a pathogen. When activated, this chimeric gene has been shown to synthesize an RNase enzyme that destroys the RNA in the cell, leading to the death of the cell. This tricky approach is rather complex, and is still under investigation. However, the general strategy to accelerate the death of infected cells has shown promise.

TRANSGENIC PLANTS WITH RESISTANCE TO FUNGAL PATHOGENS

The development of fungus-resistant plants is a key aim of plant breeding. In principle, transgenic plants harboring one or more fungal disease- resistant gene(s) can be produced to control fungal diseases, thus minimizing the use of chemical fungicides. The currently available genes to be used for producing insect-resistant and fungal disease-resistant transgenic plants include the following: genes that encode for antifungal chitinases and glucanase. These enzymes have been found in plants and microbes, and they are capable of degrading the cell walls of fungi. This is because chitin and glucan comprise major components of the cell walls of most fungi. [133, 136]

Over-expression of these hydrolytic enzymes in the plant cells is postulated to cause hyphal lysis, thereby inhibiting growth of the fungal pathogen in the transgenic plant. [133, 147] In a biotechnological breakthrough, the gene encoding a strongly antifungal endochitinase from the mycoparasitic fungus *Trichoderma harzianum* was transferred to tobacco and potato. High expression levels of the fungal gene were obtained in different plant tissues, which had no visible effect on plant growth and development. Selected transgenic lines were highly tolerant or completely resistant to the fungal foliar pathogens *Alternaria alternata*, *A. solani*, *Botrytis cinerea*, and the soil borne pathogen *Rhizoctonia solani*. [136, 146, 133]

In yet another biotechnological development, researchers at the Queensland University of Technology (QUT) in Australia have developed a genetically modified (GM) banana variety with resistance to the deadly panama disease (fusarium wilt) which is found in most of the world's banana-producing regions. Reports show that this was accomplished by taking both Cavendish and lady finger banana varieties, and inserting a single gene for this resistance into the banana genome. The gene inhibits the cell from dying when it's attacked. The GM banana variety has only been tested in greenhouses pending field trials. It is believed that the resistance generated in the glasshouse will certainly transfer through to the field. It is widely acknowledged that the development of fusarium-resistant banana varieties will provide a more permanent solution to the control of this pest, since other methods of fending off the disease have proved ineffective. In addition to developing the fusarium-resistant genetically modified banana, the QUT researchers are also trying to develop genetically modified bananas with enhanced nutritional content. Bananas are a staple food in some developing countries. The QUT project is therefore of key importance because malnutrition and deficiency of vitamin A and zinc is a major problem in Africa. [144]

TRANSGENIC PLANTS EXPRESSING BACTERIAL TOXINS

A major challenge for farmers has always been in protecting their crops from pathogens like insects, viruses, bacteria, nematodes and fungi. Pests are often kept at bay by spraying chemical pesticides on crops. Some of these chemicals have been shown to have adverse effects on the environment. [158] Crops bioengineered for pest resistance could increase yield, eliminate the use of several insecticides (now derived from fossil fuel), and reduce health risks and groundwater contamination. Transgenic insecticide producing plants have been developed by inserting a toxicity gene from a bacterium called *Bacillus thuringiensis* (Bt). The toxic gene causes a protein to be produced that is lethal to many leaf-eating insects like the maize stalk borer. [16, 36, 154]

Bacillus thuringiensis is a soil bacterium that produces insecticidal crystal proteins during sporulation. The crystals produced by this bacterium are lethal to most lepidopteran and some dipteran larvae, but several strains have been discovered to be toxic also to certain coleopteran larvae. Preparations of crystals and spores from this bacterium are used commercially as a highly selective agricultural pesticide all over the world. The Bt toxin produced by this bacterium can kill an insect if ingested. Once the gene that is responsible for producing this toxin is expressed in plants, it selectively kills insects that try to eat the plants while not having any effect on harmless insects. When the insect eats the transgenic plant leaves, it ingests the protein; and enzymes in the insect stomach or gut convert the protein into a lethal toxin that causes paralysis and death of the insect. Since the necessary enzymes that activate the Bt toxin are not found in other insects or animals, the Bt protein is harmless to them. [16, 36, 132, 137, 140]

Several transgenic crops are already available, expressing different Bt proteins. These include corn, cotton, potato, soya bean, and some tomato plants. Field trials have shown that they are indeed resistant to the corresponding insect pests. However, a challenge in the use of Bt is to avoid generating excessive numbers of insect pests that are Bt resistant. While the presence of a few insects that are naturally resistant to Bt is always inevitable, the concern is that widespread use of Bt will kill all those insects that are not naturally resistant, leaving only the resistant ones to multiply. This could create a population of insects that were resistant to the pesticide as a result of their constant exposure to the toxin. As a matter of fact, insects are already showing resistance to BT toxin. This means that after a few generations, nearly all of the insect pests targeted by Bt would be Bt resistant. Well, the issue of Bt resistance is a big concern and requires the adaptation of integrated pest management strategies. In addition, many scientists advise that whenever a Bt crop is planted, farmers should establish a certain percentage of non-Bt plants to reduce the risk of the development of insect resistance. The best way to go about it would be to require seed companies to sell a correct mixture of Bt and non-Bt seeds. [16, 36, 132, 137, 140]

TRANSGENIC PLANTS WITH RESISTANCE TO
NEMATODE PATHOGENS

Plant parasitic nematodes infect roots of crops causing damage which reduce their capacity to produce high yields. In particular, root-knot nematodes of the *Meloidogyne* spp. which are wide spread in tropical and subtropical soils are notorious pests of many staple food crops. The interaction of plant parasitic nematodes with their host plants, certainly in the case of sedentary parasites, is a complex multi-step process. Plant biotechnologists have established ways of trying to interfere with this process with the aim of developing transgenic plants that are resistant to nematodes. For example the nematode feeding site can be destroyed or inhibited in its function, resulting in the starvation of the nematode. [138, 147, 151]

The survival of root-knot nematodes is sustained by the formation of giant cells or syncytia in the host plant tissue, upon which they feed. The development of giant cells can be prevented by inserting a very strong cytotoxic gene (ribonuclease barnase) to be expressed in the feeding cells. The expression of this gene in the roots of the plant has been found to degrade the giant cells, thereby destroying their function of feeding the nematode. This has been achieved by using promoters that are highly active in feeding cells but also to a lower extent in some other plant organs. This background expression in other plant cells is of course harmful for the plant, but can be inhibited or countered by expression of barstar (an inhibitor of barnase) in these other cells. [138, 147, 151]

In yet another biotechnological development, nematode resistant rice and potato have been developed by the introduction of sunflower and modified rice cystatin genes. The gene product is confined to the plant roots by linking the gene to a root-specific promoter. This ensures the gene product is not present in the edible plant parts and will only be present in the plant roots where it can specifically target nematodes. Cystatins are naturally occurring proteinase inhibitors used by plants as a defense against insects and pests. For example, rice cystatin is naturally found in rice grain and forms part of the human diet. They work against nematodes by preventing them from digesting their food from plant roots properly and so they are unable to reach their egg laying size. A key feature is that cystatins are effective against a wide range of nematodes and so can protect banana from different combinations of pest species that can occur in plantations. Field trials in the UK have shown that these transgenic potato lines have developed up to 70% resistance against nematodes. However, this so called new crop-line is yet to gain approval for commercialization. [145, 147]

TRANSGENIC PLANTS SHOWING COAT PROTEIN-MEDIATED RESISTANCE AGAINST VIRUS INFECTION

Ref.: [34, 134, 142]

There are a large number of viruses that infect plants causing serious losses in production, both in terms of yield and quality. The vast majority of viruses that infect plants are RNA viruses. This means that the nucleic acid carried in the virus particle, the viral genome, is comprised of RNA. In addition, there are a small number of plant DNA viruses, including Cauliflower Mosaic Virus (CaMV).

Plant viruses are fairly simple creatures; they are made up of a protein shell or coat which surrounds the RNA or DNA genome. The viral shell is made up of one or two proteins, and these proteins assemble to form the coat. These proteins are known as coat proteins. When a virus enters a plant cell, the RNA is released from the protein coat into the cytoplasm of the plant cell. This RNA is then used as template to make RNA and proteins encoded by the virus. The genome of the virus is replicated and new coat proteins are synthesized. This leads to the assembly of new virus particles in the infected cells of the plant.

Virus transmission to other plants can occur in a number of ways:

1. Mechanically by physical touching of infected and healthy plants. For example, in greenhouses some viruses are spread by using the same tools, such as pruners, which carry virus from infected plants to those that are healthy. Good sanitary practices can reduce the spread of viruses.

2. By biological vectors such as aphids or other insects that feed on an infected plant, ingest some virus particles,

then move to a healthy plant and transmit the virus through the saliva when feeding.

3. Some viruses are carried in the seed and then passed from one generation to another within the seed. Plant viruses can also be transmitted through grafting of infected tissue onto healthy tissue.

Many viruses produce a protein that helps the virus move through the plant from the original site of infection. This movement protein enlarges the small passageways between cells (the plasmodesmata) so that virus particles can get through, allowing them to spread systemically from cell to cell. There are a number of strategies in biotechnology that can be used to produce transgenic plants with resistance to viruses: The principle method is to identify single genes that provide resistance to a particular virus. For example, the gene for resistance to a tobacco mosaic virus (TMV) has been cloned. Transgenic tobacco plants have been engineered to express the coat protein of tobacco mosaic virus (TMV). As a result the plant cells express the protein, and this makes them "immune" to infection by complete TMV virus, which would cause the plant leaves to wither and die. The immunity occurs because the TMV virus does not infect a cell that is already TMV infected. The presence of the single harmless protein in all the plant cells tricks the TMV viruses into behaving as if the cells of the plant were already infected.

When transgenic tobacco plants that express the coat protein of TMV were exposed to the virus, they were found to be resistant to infection by the virus. The inoculated transgenic plants did not develop symptoms, while the control, non-transformed plants did become infected. However, the transgenic plants expressing the TMV coat protein were not completely immune; they would become infected if inoculated with a high dose of the virus, but it took longer for symptoms to develop and the symptoms were less severe. This protection works not only against the virus from which the coat protein was isolated, but also against closely related viruses.

Having been shown to provide protection against TMV in tobacco, the same strategy (expression of the virus coat protein in transgenic plants) was then applied to other plant viruses. This approach has now been shown to be effective in providing resistance against a large number of different plant viruses. This is referred to as coat protein-mediated resistance. Another vegetable crop that has been developed with resistance to a virus using coat protein mediated resistance is potato. Resistance to potato leafroll virus and potato virus Y has been found to be successful.

Other methods have also been tested to develop plants with resistance to viruses. When the replicase of the virus is expressed in transgenic plants, this also provides resistance to the virus from which the replicase was cloned. Expression of coat protein and replicase are the best examples of what is called pathogen-derived resistance, where part of the virus is used to make plants resistant to that pathogen. The methods include:

1. Using antisense RNA against the viral genome.

2. Expressing an enzyme from mammals that modifies the viral RNA thereby inhibiting its replication.

3. Expressing ribozymes (short RNA molecules) that are able to bind to specific RNAs in the cell, in this case the viral RNA, and break it into smaller pieces.

   Targeting the movement of viruses from one plant cell to another. This intercellular movement of viruses requires that the plasmodesmata, the pores between cells, are enlarged. Many viruses produce a movement protein that is responsible for this modification of plasmodesmata. From detailed studies on viruses like TMV, it was shown that some strains of virus were unable to move from cell to cell and produced very limited disease symptoms. The reason these viruses could not move was because they have a defective movement protein, one that prevented the plasmodesmata from enlarging.

When this defective movement protein was expressed in transgenic plants, it was found to provide resistance against viruses, not by preventing the infection of plant cells but by preventing the spread of virus from cell to cell.

All of these methods have been shown to work to some degree, but none have been developed into a commercial application to date.

Virus-Resistant Papaya

The papaya plant, which originally came from Central and South America, is popularly grown throughout the tropics. Papaya ringspot virus (PRSV) is a virus that infects papaya and is a big hindrance to production of this tropical fruit. The virus is transmitted by insects such as aphids, but insecticide treatments against aphids are not effective. Over the last ten years researchers at a number of institutions have been developing transgenic papaya that express the coat protein of PRSV using the same strategy previously outlined. This has led to the release of a transgenic papaya with resistance to PRSV which is now being planted in Hawaii. This product has been approved by USDA, EPA and FDA. [34, 142]

LITERATURE CITED

[132]Anzai, H. Yoneyama, K. & Yamaguchi, I (1989). Transgenic tobacco resistant to a bacterial disease by the detoxification of a pathogenic toxin. Molecular and general Genetics 219:492–494.

[133]Aparna Islam (2006). Fungus Resistant Transgenic Plants: Strategies, Progress and Lesson Learnt. Plant Tissue Cult. And Biotech. 16 (2): 117–138, 2006 (December).

[134]Beachy, R. N. Loesch-fries, S. & Tumer, N. E. (1990). Coat protein mediated resistance against virus infection. Annual review of Phytopathology 28:451–474.

[135]Botterman, J. & Leemans, J. (1988). Engineering herbicide resistance in plants. Trends in Genetics 4: 219–222.

[136]Brogile, K. chet, I., Holliday, M., Cressman, R., Biddle, P., Knowlton, S., Jerry Mauvias, C. & Brogile, R. 1991. Transgenic plants with enhanced resistance to the fungal pathogen *Rhizoctonia solani*. Science 254: 1194–1197.

[137]Brunke, K. J. & Meeusen, R. L. 1991. Insect control with genetically engineered crops. Trends in Biotechnology 9: 197–200.

[138]Cai, D., Kleine, M., Kifle, S., Harloff, H.-J., Sandal, N.N., Marcker, K.A., Klein-Lankhorst, R.M., Salentijn, E.M.J., Lange, W., Stiekema, W.J., Wyss, U., Grundler, F.M.W., and Jung, C. (1997) Positional cloning of a gene for nematode resistance in sugar beet. Science 275:832–834

[139]Erik A. van der Biesen (2001). Quest for antimicrobial genes for engineering disease-resistant crops. Trends in Plant Sc. 6:89–91.

[140]Fedoroff, Nina V. Fedoroff, N. V. (2003). "Prehistoric GM corn". *Science* [302]: 1158–1159. doi:10.1126/science.1092042. PMID 14615520

[141]Fraley Robert (1992). Sustaining the Food Supply. Biotechnology. Vol 10. January 1992. pp 40–43.

[142]Gonsalves, D. (1998). Control of papaya ringspot virus in papaya: A case study, Annual Review of Phytopathology 36:415–437.

[143]Halford NCordell, Halford. N (2003). *Genetically Modified Crops.* Imperial College Press. ISBN 1-86094-353-5 .

[16]Hofte H, de Greve H, Seurinck J., et al (December, 1986). "Structural and functional analysis of a cloned delta endotoxin of Bacillus thuringiensis Berliner 1715". Eur. J. Biochem. 161 (2):273–280. doi:10.1111/j.1432-1033.1986.tb10443.x. PMID 3023091. http://www.blackwell-synergy.com/openurl?genre=article&sid=nlm:pubmed&issn=0014-2956&date=1986&volume=161&issue=2&spage=273.

[144]Jayne Margetts (2008). Australian Broadcasting Corporation Australian Broadcasting Corporation: *Summary posted by Meridian on 10/28/2008.*

[21]Leighton Jones, publication manager: BMJ. 1999 February 27; 318 (7183): 581–584. Copyright © 1999, British Medical Journal - Science, medicine, and the future - Genetically modified foods.

[145]Liu B, Hibbard JK, Urwin PE, Atkinson HJ. The production of synthetic chemodisruptive peptides in planta disrupts the establishment of cyst nematodes. *Plant Biotechnol J.* 2005, 3(5) 487–96.

[146]Matteo Lorito, Sheridan L. Woo, Irene Garcia Fernandez, Gabriella Colucci, Gary E. Harman, José A. Pintor-Toro,

Edgardo Filippone, Simona Muccifora, Christopher B. Lawrence, Astolfo Zoina, Sadik Tuzun & Felice Scala (1997) . Genes from mycoparasitic fungi as a source for improving plant resistance to fungal pathogens. Communicated by R. James Cook, Agricultural Research Center, Pullman, WA (received for review November 24, 1997).

[147]Mauch F and Staehelin LA 1989. Functional implications of the subcellular localization of ethylene-induced chitinase and beta-1,3-glucanase in bean leaves. Plant Cell. 1:447–457.

[148]Milligan, S. B., Bodeau, J., Yaghoobi, J., Kaloshian, I., Zabel, P., and Williamson, V. M. (1998). The rootknot nematode resistance gene Mi from tomato is a member of the leucine zipper, nucleotide binding, leucine-rich repeat family of plant genes. Plant Cell 10, 1307–1319.

[149]Mittler R., Del pozo O., Meisel L., Lam E. (1997). Pathogen induced programmed cell death in plants: a possible defense mechanism. Dev. Genetics 21:279–289.

[150]Naisbitt John and Patricia Aburdene. Megatrends 2000. Pan Books. 1990.

[151]Narayanan, R. A., R. Atz, R. denny, N.D. Young, and D. A. Somers (1999) Expression of Soybean Cyst Nematode resistance in Transgenic Hairy Roots of soyabean. Crop Sci. 39, 1680–1686.

[152]Reddy, K. N., and K. Whiting (2000). Weed control and economic comparisons of glyphosate-resistant, sulfon lurca-tolerant, and conventional soy bean (Glycine max) systems. Weed Technology 14:204–211 Regulation (EC) 1829/2003 on GM food and feed – http://www.food.gov.uk/gmfoods/gm/evaluating

[153]Ribas Alessandra Ferreira, Luiz Filipe Protasio Pereira, and Luiz Gonzaga E. Vieira (2006). Genetic transformation of coffee. Physiol. vol.18 no.1 Londrina Jan./Mar. 2006.

[154]Snow AA, Andow DA, Gepts P, Hallermann EM, Power A, et al., (2004). Genetically engineered organisms and the environment: current status and recommendations: http://www.esa.org/pao/esaPosition/Papers/geo_position.htm - 2009.

[34]Snow, A. A., D. A. Andow, P. Gepts, E. M. Hallerman, A. Power, J. M. Tiedje, and L. L. Wolfenbarger. (2005). Genetically engineered organisms and the environment: Current status and recommendations. Ecological Applications, July 16, 2004.

[36]*Vaeck M, Reynaerts A, Hofte A, et al. (1987). "Transgenic plants protected from insect attack". Nature 328: 33–7:* doi:10.1038/328033a0. http://www.nature.com/nature/journal/v328/n6125/abs/328033a0.html.

CHAPTER VII

Novel Applications of Transgenic Technology

GENE MANIPULATION TO REDUCE ALLERGENECITY OF FOODS

Hay fever and allergenic asthma triggered by pollen from sources such as grasses, maize and rye are a common occurrence. Apart from allergies that result from exposure to such pollen, peanuts, soybean, and various fruits are also notorious for causing allergic reactions. For example, many people suffer from apple allergy. Related fruits such as pear, cherry and peach have also been reported to cause allergy. [170, 171, 202] Accompanying this is the rising concern by the public regarding the creation of new allergens in GM foods. [179]

Reduction of allergens in food can be accomplished using two ways: (1) through the development of hypoallergenic primary material; and (2) through the destruction or elimination of allergens or allergenic epitopes by food processing. The development of hypoallergenic primary material can be done through conventional plant breeding methods, or genetic modification. Most apple-allergic patients avoid eating the fruit, thus missing out on a food source of high nutritional value; "An apple a day keeps the doctor away" is a common saying. It is worthwhile, therefore, to develop strategies that can prevent production of allergens by such essential foods. [170]

are various technologies that can be used to reduce the allergenicity in primary food products and this include: (a) selection of low allergenic cultivars from existing biodiversity of a given crop; (b) breeding using characterized genotypes and genetic markers for low allergenicity; (c) genetic modification to silence an allergen gene. [170, 175] Plant genetic engineering has the potential to both introduce new allergenic proteins into foods and remove established allergens. It has been established that genetic engineering can make a food less allergenic. Soybean, a known cause of allergies in humans, has been genetically engineered to shut down the gene that codes for the protein believed to cause most soybean allergies, using a transgene-induced gene silencing technology. This novel approach to reducing allergies should add nutritional value to crops. [175, 179]

GENE MANIPULATION TO IMPROVE THE PROCESS OF PHYTOREMEDIATION

Widespread contamination of the environment caused by the manufacture, testing, and disposal of explosives is becoming a matter of great concern. The enormous growth of industrialization and the use of numerous pesticides, dyestuffs, explosives, and pharmaceuticals have resulted in serious pollution of the environment in many parts of the world. Elemental pollutants are particularly difficult to remediate from the soil, water and air because, unlike organic pollutants that can be degraded to harmless small molecules, toxic elements such as mercury, arsenic, cadmium, lead copper and zinc are non-degradable hence remain in the ecosystem. [169, 181] It is a common observation that through seepage, these residual contaminants find their way into surface water and groundwater, and subsequently enter the food chain through plants grown on such soils. [183]

The occurrence and accumulation of endocrine disrupting chemicals (EDCs) such as some pesticides in the environment has been found to be detrimental to wildlife, including birds and fish. [177] Researchers have implicated EDCs in the following body malfunctions: reproductive infertility or decreased fertility; sexual under-development; altered or reduced sexual behavior; attention deficit/hypersensitivity disorder; altered thyroid and adrenal cortical function. The EDCs have also been linked to increase in certain cancers and birth defects in animals. [203]

Plants have many natural properties that make them ideally suited to clean up polluted soil, water, and air, in a process called phytoremediation. The potential use of plants to remediate contaminated soil and groundwater has recently received a great deal of interest. The science of phytoremediation arose from the study of heavy metal tolerance in plants in late 1980s. [166] The discovery of hyper-accumulator plants, which contain levels of heavy metals that would be highly toxic to other plants, prompted the idea of using certain plant species to extract metals from the soil and, in the process clean up soil for other less tolerant plants. Scientists also found that certain plants could degrade organic contaminants by absorbing them from the soil and metabolizing them into less harmful chemicals. [174]

More recently, genetic engineers and other scientists have applied a type of plants known as phreatophytes to remediation of contaminated groundwater. The term phreatophtes literally means water-loving plants such as hybrid polar trees. Phreatophytes have the ability to adapt to desert conditions by developing long root systems to draw water from deep underground near the water table. Some roots of such plants have been recorded at 80ft. [166, 172]

Apart from phytoremediation, other remedial processes that can be used to clean up such environments include: (a) vitrification – a waste disposal method for mobilizing and encapsulating radioactive and other types of hazardous materials. In vitrification process, high temperature is employed to melt the material into a liquid, which on cooling, transform to an amorphous, glass-like solid and permanently captures the waste; (b) bioremediation – the use of biological agents, such as bacteria or plants, to break down and thereby detoxify dangerous chemicals in the environment; (c) earth-swap or solifluction – gradual movement of wet soil and so forth down a slope; (d) soil flushing – involves the flooding of the soil with a flushing solution, which may be acidic, basic, or containing surfactants and the subsequent removal of the leachate via shallow wells or subsurface drains.

In many cases, phytoremediation may have a cost advantage over other treatment technologies because it relies on the use of the natural growth processes of plants, and often requires a relatively small investment in both capital and maintenance costs. [167] Thus modern biotechnology researchers are leaning towards optimization of the use of phytoremediation because it is a cost-effective and efficient means of cleaning up the environment.

There are various types of phytoremediation: [169, 185, 198]

1. Phytoextraction – the use of plants, mostly roots, to re-
    move contaminants from soils.

2. Phytovolatization – the enhancement of the volatiliza-
    tion of chemical contaminants by plants. Phreatophytes
    have been found to remove contaminants from the envi-
    ronment through metabolism, converting it all the way
    to normal end points such as carbon dioxide and salts.
    [165]

3. Phytofiltration – refers to the use of plant roots (rhizo-
    filtration) or seedlings (blastofiltration) to absorb or
    adsorb contaminants from flowing water. Some plant
    tissues have the ability accumulate contaminants.
    Poplar trees have demonstrated the ability to uptake
    and store heavy metals in intracellular root spaces, and
    to translocate these compounds to shoots and leaves.
    [176] However, a point of caution is that as such plants
    lose their leaves or die, the organic matter needs to be
    collected and transported to an appropriate waste facil-
    ity, so that the contaminant is not re-introduced into the
    environment.

4. Phytostabilization – the use of plants to transform soil
    metals to less toxic forms, that is to say, to reduce the
    bioavailability of the pollutant in the environment. This
    process does not however, remove the metal from the
    environment.

5. Phytodegradation – the use of plants to degrade organic
    contaminants. Phreatophyte roots may break down con-
    taminants in soil through the effect of the enzyme deha-
    logenase, or by producing root exudates that transforms
    or mineralizes contaminants. [197]

6. Rhizosphere bioremediation – the use of plant roots in conjunction with their rhizosphere microorganisms to remediate soil contaminants. Some plants assist in the breakdown of contaminants in soil through enhancement of microbial activity in the rhizosphere. Plant roots provide passive aeration, serve as a nutrient source for microbes, and also serve to draw water to the surface. [182]

Phytoremediation is a relatively new technology, and it is being applied to a variety of contaminants, including environments polluted with oil spillages, pesticides, solvents, explosives, radionuclides, heavy metals, and land chelates. [167]

Some plants which grow on metalliferous soils have developed the ability to accumulate massive amounts of the indigenous metals in their tissues without exhibiting symptoms of toxicity. [156] Chaney (1983) [157] was the first to suggest using these "hyperaccumulators" for the phytoremediation of metal-polluted sites. However, many of these so-called hyperaccumulators were found to have limited potential in this area due to slow growth, which limit the speed of metal removal. [159]

Many systems have exploited the use of Poplar and Cottonwood trees because they are fast-growing and have a wide geographic distribution. [166] Poplar trees may provide a hydraulic control of aqueous contaminants, containing subsurface water through uptake, thus decreasing the tendency of surface contaminants to move toward groundwater. Poplars have been shown to transpire from 50 to 300 gallons of water per day based on the prevailing environmental conditions. [158] Examples of other types of vegetation used in phytoremediation of surface soils include Sunflower, Indian mustard, and grasses such as ryegrass and prairie grasses. [167]

The efficiency of phytoremediation or hyper-accumulation of contaminants can be achieved, either through using genetic approaches, or by improving the plants' biomass and environmental requirements. Increasing the plant biomass is achieved through increasing nutrients. Genetic engineering approaches involve modifying the plants metabolism and the addition of new phenotypic and genotypic characteristics to the plant. [162, 169] Other researchers have recorded success in enhancing the plants' natural bioremediation products through the over expression of biosynthetic genes using a strong promoter. [187]

Alternately, researchers used tissue culture to select the desired genes, such as those coding for products with enhanced ability to assimilate heavy metals. The new plant varieties containing the new traits are then regenerated. Over the years, tissue culture technique has been extended to incorporate recombinant DNA technology, whereby the desired gene is isolated and introduced into the plant cells. This is done by physical means (electroporation or via high-velocity microprojectiles shot into the cell), or by using a biological vector, *Agrobacterium tumefasciens*. Once the desired gene is incorporated into the plant cells, the transformed cells are then selected using selectable markers such as antibiotics, and the cells grown in tissue culture are used to regenerate whole plants for subsequent breeding.

Metal hyperaccumulation plants and microorganisms with unique abilities to tolerate, accumulate and detoxify metals, represent an important reservoir of unique genes. [160] Such genes could be transferred to fast-growing plant species for enhanced phytoremediation. [163] Certain soil bacteria are known to have biodegrading capability. Scientists in England successfully introduced pentaerythritol tetranitrate reductase, the bacterial enzyme initiating degradation of explosive residues, into plants, and the transgenic plants so created were used for bioremediation of contaminated soils. [191, 205] Such an application of biotechnology has great promise for cleaning the environment. Generally, genetic modification to improve the process of phytoremediation has recorded great success, although it is still in its research and developmental phase, and has a long way to go before such plants are commercialized. [169]

GENETIC DECAFFEINATION OF COFFEE

Tea (*Camelia sinensis*) and coffee (*Coffea arabica*) provide some of the most widely used beverages in the world. As much as people like to have tea or coffee, some of them would like to have little or no intake of caffeine, an important stimulant in both tea leaves and coffee beans. [179] Low caffeine and decaffeinated coffee represents around 10% of the world's coffee sales. The demand for decaffeinated coffee is growing globally, due to the possible adverse health effects of caffeine. Caffeine is a mild stimulant and can cause nervousness, irritability, and insomnia. Caffeine has also been found to trigger heart palpitations and increase blood pressure in sensitive individuals. [179, 180]

Decaffeination is the act of removing caffeine from coffee beans, cocoa, mate, tea leaves and other caffeine-containing materials. In the case of coffee, the decaffeination process is usually performed on unroasted (green) beans to remove the stimulant before the coffee beans are roasted. Various methods for selectively extracting caffeine have been used over the last century: [161, 173]

(a). Water extraction - Hot water extracts both flavor ingredients and caffeine from green coffee beans. If the extract is passed through activated charcoal, most of the caffeine is removed by adsorption. Soaking the original beans in the decaffeinated extract then restores most of their flavor.

(b.) Organic solvent extraction – the coffee beans are rinsed with a solvent that contains as much of the chemical composition of coffee as possible without also containing the caffeine in a soluble form. Low toxicity solvents such as ethyl acetate are used. Although it has been found to be moderately toxic, coffee makers have touted ethyl acetate as "natural" because it was present in fruit.

(c.)   Supercritical carbon dioxide extraction – involves a process whereby supercritical carbon dioxide is forced through green coffee beans. Supercritical carbon dioxide is a fluid which has both gas-like and liquid-like properties. Its gas-like behavior allows it to penetrate deep into the beans, dissolving the caffeine present. Coffee manufacturers then recover the caffeine and resell it for use in soft drinks and medicines.

Coffee contains over 400 chemicals important to the taste and aroma of the final drink. This makes it challenging to remove only caffeine while leaving the other chemicals at their original concentrations. The commercial processes of decaffeination currently available are not only expensive, but they leave certain chemical residues, and may also lead to loss of flavor and aroma of coffee. [161, 173]

Methods of genetically decaffeinating coffee have been tried with remarkable success. Caffeine is produced in coffee plant cells from a natural plant chemical called xanthosine. Caffeine synthase is the enzyme that catalyses the final two steps in the caffeine biosynthesis pathway. Several researchers have developed transgenic plants with suppressed caffeine synthesis [180, 188, 189] This has been accomplished using the antisense strategy or gene silencing, paving the way for creating tea and coffee plants that are naturally deficient in caffeine. Generally, proteins are produced from genes in living cells via complementary molecules called messenger RNA (mRNA). This mRNA, which is a mirror image of the gene's DNA, is "read" to produce a protein, but RNA interference through antisense strategy or gene silencing switches off this process.

Researchers have genetically engineered coffee plants that have 70% less caffeine than usual in their leaves. It is hoped that this so-called "new" coffee could yield low-caffeine beans, on maturation, which will take approximately 3 or 4 yrs [188, 190] and these transgenic beans could rival industrial decaffeination if they gain public approval. [179, 193]

GENETICALLY MODIFIED OIL

The nutritional and industrial value of many oils is determined by its fatty acid profile, including the level of desaturation. Omega-3 fatty acids (unsaturated fatty acids) are used by the human body to make anti-inflammatory and anti-thrombotic substances, while omega-6 (saturated fatty acids) are made into substances that promote inflammation in the body and thrombosis. The beneficial effects of omega-3 fatty acids include reducing heart disease, reducing circulating cholesterol levels and suppressing inflammation in humans and animals. Therefore not only is the presence of both the saturated and unsaturated fatty acids in the diet necessary for health, but also the ratio of unsaturated to the saturated fatty acids is critical in achieving an appropriate balance of fatty acid derived functions in the human body. [192, 194] The conventional approach to fatty acid modification in plants has been to explore natural or induced mutations occurring in the same plant species or close relative, but this is relatively slow and labor-consuming. [192]

There have been several transgenic breeding approaches aimed at improving the ratio of unsaturated to the saturated fatty acids in plants. The fatty acid composition in several oil seed varieties has been significantly modified following the introduction of transgenes, using *Agrobacterium*-mediated transformation system. The transgenes include those coding for the enzymes which play important roles in the formation of fatty acids and the incorporation of the fatty acids into triacylglycerols. The transgenes can be isolated from related plant species or other plants, yeast, bacteria and some mammals. [192] An alternative approach to decrease the level of saturated fatty acids within plants is by regulating the expression of a number of genes including acyltransferases, ketoacyl-acyl carrier protein synthatases (KAS), desaturases, and thioesterases. [164, 186]

Other biotechnological exploits regarding genetic modification of oil are also ongoing. In a bid to reduce dependability on foreign countries for the production of tropical oils such as palm and coconut, scientists have come up with very interesting innovations. Currently, one oilseed species can be made to produce a number of new food-grade oils, through genetic manipulations. Oilseed rape can now be modified to produce oils with wide ranging characteristics through selective modification of the length and degree of saturation of the fatty acids produced. Fatty acids such as lauric acid, typical of tropical vegetable oils, can now be produced in temperate oilseed crops, although this is still under trial. [21, 168]

In yet another biotechnological development, researchers have shown that enzymes can be inserted into sunflowers so that they can convert their own sunflower oil into palm or coconut oils. [21, 168]

FROM PLANTS TO PLASTICS

Plastics have become synonymous with modern life, but improperly disposed plastic materials are a significant source of environmental pollution, potentially harming life. The plastic sheets or bags do not allow water and air to seep into the earth, thereby reducing the fertility and aeration of the soil, depleting underground water and harming animal life. When ingested by livestock, plastics grossly interfere with digestion and uptake of nutrients. The same is true of marine animals like whales, dolphins, turtles and seabirds. Most of today's plastics and synthetic polymers are produced as by-products from non renewable energy resources such as petroleum, coal and natural gas. The central problem with plastics from hydrocarbons is that they are not biodegradable because of their long chain polymer molecules. "Biodegradable" means that a substance can naturally decompose with the aid of microbial organisms and will not persist in the environment beyond a certain time period. [200]

A type of sturdy and hard plastic is made with a molecule known as Bisphenol A (BPA), which like many residual chemical contaminants is now detectable in most people's blood streams. BPA is associated with some cancers, early onset of puberty in children, obesity and even attention deficit disorder. To overcome this problem, biodegradable plastics have been developed which are made from renewable resources, such as plants. [199, 200, 204]

Biodegradable plastics can be developed using various techniques: [199, 200, 204]

1. Secondary compounds that are naturally produced by plants. Scientists have been able to produce plastics from corn, wheat and soybeans in the form of cellulose, starch, collagen, casein, soy protein polyesters and triglycerides. For example, most cereal crops and tubers

contain vast storage of starch, a natural bio-polymer which is readily convertible into plastic. However, basic starch is water soluble which may limit its use. But with the help of several microorganisms, it can be transformed into lactic acid, a basic unit that can polymerize to give polylactides (PLA) for use in a variety of products. Already, owing to its easy decomposability, this product has been successfully used for drug delivery and implants in the medical world.

2.   Biodegradable natural plastics can also be manufactured from bacteria. Within bacterial cells, granules called polyhydroxyalkanoate (PHA) are stored. Bacteria can be easily grown in bulk culture and the plastic harvested.

3.   Transgenic plants that can make and store large quantities of plastics in the cell organelles called vacuole.

Genetic engineering has been deployed as a novel biodegradable plastic generating tool. Corn plants containing the bacterial gene for producing PHA have been shown to produce the same variety of plastics within their cells. [200, 204] Other reports show that a decade ago, scientists at Agracetus Co. in Wisconsin found a way to insert a gene into a cotton plant so that the plant expresses a polyester polymer inside the cotton fiber, thereby automatically producing a cotton-polyester blend. Because this eliminates the manufacture of plastics it reduces environmental pollution. [201]

In yet another biotechnological development, a gene has been expressed in potatoes; that makes a plastic polymer which is biodegradable. This fiber can be used to manufacture items like grocery bags that will be totally recycled after use because of their biodegradability. However, reports show that this development is at the experimental stage and it will take a while before it is approved for commercialization. [201]

GENETICALLY MODIFIED RICE TO PREVENT HYPERTENSION

Hypertension or high blood pressure, as it is more commonly known, is regarded as a silent killer. It is a disease of modern age. The fast pace of life and the mental and physical pressures caused by the increasingly industrialized and metropolitan environments have a role to play in the rise in blood pressure. Blood pressure is measured with an instrument called sphygmomanometer in millimeters of mercury. The highest pressure reached during each heart beat is called systolic pressure, and the lowest between two beats is known as diastolic pressure. Most young adults have blood pressure around 120/80 mm Hg.

The level of blood pressure increases with the advancing age even in absence of other causative factors. In women tie blood pressure increases at the time of menopause and higher level persists in majority of them in later years. The blood pressure goes higher in those who have family history of hypertension and in those who suffer from type 2 diabetes. It also goes higher in those who have suffered from acute nephritis as a complication of sore throat in early childhood. The blood pressure is considered high if the level is above normal blood pressure (138/85-89 mm Hg). The level of blood pressure is divided into 3 stages.

Hypertension	(140-159/90 to 99 mm Hg.)	- Stage I
Hypertension	(160-170/100–109 mm Hg.)	-Stage II
Hypertension	(190-230/110-119 mm Hg.)	-Stage III

Above Stage III level it is called malignant hypertension. Risks of complications of hypertension increases with persistent untreated hypertension and when associated with obesity, increased lipids level of blood, family history of hypertension, post menopausal status in women, smoking and presence of type 2 diabetes. The target organs are heart hypertrophy (enlargement of ventricles) - (cardio vascular diseases) kidney leading to its failure and brain causing stroke. [195, 196]

Treatment is by drug therapy. Life style modification can also provide a cost effective approach to prevention of complications or management of hypertension. The required life style modification is simple but difficult to observe by some hypertensive people. This includes weight loss and sodium (common salt intake) restriction. Over weight (obesity) increases the risk. Increased waist circumference is a weight related risk. Waist circumference of more than 88 cm (35 inches) in women and 100 cm (40 inches) or more in man is considered health risk for hypertension. Losing weight can lower pressure. Many persons with mild hypertension will be able to control their blood pressure with weight loss alone and those with higher blood pressure will require less medication for their blood pressure control. Studies have shown that reducing dietary salt intake will lower the blood pressure in those who have higher blood pressure. They should restrict the salt intake to daily 2.5g. They should avoid salt containing food i.e. pickles, lemon juice, processed foods (to prefer to eat freshly prepared food). Salt restriction help to lose weight. [195, 196]

Complimentary to salt restriction is increased potassium intake. It is derived from fresh fruits or vegetables. Supplement of potassium salt intake lowers the blood pressure. Potassium chloride or citrate in dose of 3 to 4 gm. per day will lower the blood pressure. It has been found that increased intake of vegetables or/and fresh fruits will provide substantive levels of potassium enough to lower the blood pressure. Increased calcium and magnesium intake will also help to lower the blood pressure because low intake of calcium and magnesium has been correlated with higher blood pressure. Excessive intake of alcohol is also associated with the development of blood pressure. Caffeine intake (coffee beverages) increases blood pressure on intake but the body develops caffeine adjustment in the long run with hardly any rise in blood pressure. Thus simple and essential nutritional requirement is a means of non-drug treatment to control hypertension. [195, 196]

A third alternative to preventing hypertension lies in eating a "new" kind of genetically modified rice. Researchers from Japan have developed transgenic rice that is capable of accumulating significant levels of the anti-hypertensive proteins gamma-aminobutyric acid (GABA) and nicotianamine (NA). The researchers from Shimame University developed rice lines that express increased levels of the four-carbon amino acid GABA. GABA, an inhibitory neurotransmitter in mammalian central nervous system, has been shown to lower blood pressure in animals. A modified form of the gene that encodes for glutamate decarboxylase (GAD), under the control of the rice glutelin promoter (GluB-1), was introduced to rice cells via Agrobacterium-mediated transformation. [178]

In yet another biotechnological development, other researchers have developed rice plants that produce the ACE inhibitor nicotianamine (NA). ACE or angiotensin I-converting enzyme is a key enzyme in hypertension and studies have shown that inhibition of its activity leads to reduced blood pressure. ACE inhibitors are widely accepted as the drugs of first choice for patients with hypertension and congestive heart failure. The scientists found that the ACE inhibitory activity of the transgenic rice-derived NA is very strong, even when compared with commercially available antihypertensive peptides. To minimize public anxiety over the GM rice, the selectable marker genes for antibiotic resistance were removed using the Cre/loxP DNA excision system. [178]

In an analogous way other scientists from Japan have also developed a "new" rice to combat hay fever. The new strain of rice contains a gene that produces the allergy-causing protein, a Farm Ministry official said on Friday. The researchers believe that, "eating this "new" rice helps mute the reaction of the body's immune system," such an effect is similar to other allergy treatments which entail releasing a small amount of allergen into the body to allow resistance to build up. [155]

MALE STERILE PLANTS

Self-fertilization has to be prevented during production of hybrid seeds. This can be done by manually removing the anthers but this is very labor intensive. A much more practical approach is to use male sterile plants. Genetic engineering has been successfully used to introduce male sterility in several crop species such as oilseed rape, tobacco, and corn. The production of pollen is prevented by destruction of the tapetum, a cell layer which is necessary to nurse the pollen forming cells. This destruction is accomplished by expressing a cytotoxic gene specifically in the tapetum layer. A cytotoxin gene that has been used is the ribonuclease barnase that degrades all RNA inside the cell. A construct molecule with this gene fused to a tapetum specific promoter is introduced into the plants: all cells of this plant will contain the gene but because of the specific promoter only the developing tapetum cells will die. Consequently no pollen is formed and the plant is male sterile, but otherwise completely normal. Cross fertilization with pollen from a male parent that expresses barstar (an inhibitor of barnase) will restore fertility and fruit formation in the next generation. [184]

LITERATURE CITED

[155]Allergies. Feb 04, 2005. Available from: http://www.health.am/ ab/more/japan_says_gm_rice_could_help_combat_hay-fever/

[156]Baker, A. J. and R. R. brooks (1989). "Terrestrial higher plants which hyper-accumulate metal elements-a review of their distribution, ecology and phytochemistry." Biorecovery 1:81–126.

[157]Chaney, R. L. (1983). Plant uptake of inorganic waste consti-tutes land treatment of harzadous eastes. P. B. M. J. F. Pan, and J. M. Kla. Park ridge, NJ, Noyes Datta Corp.

[158]Chapell, J. (1997). Phtoremediation of TCE in Groundwater using Populus, U. S. EPA Technology Innovation Office.

[159]Cunningham, S. D. & D. W. Ow (1996). "Promises and Prospects of Phytoremediation." Plant Physiol. 110. 715–719.

[160]Danika, L. LeDucNorman T (2005). Phytoremediation of toxic trace elements in soil water. J. Ind. Microbiol. Biotechnol. 32:514–520.

[161]David, K (2002) Fun without the buzz: decaffeination process and issues. Available from: http://www.virtualcoffee.com/ sept_2002/decafe.html (Accessed Dec. 2005).

[162]Davidson (2005). Risk mitigation of genetically modified bacteria and plants designed for bioremediation; J. Ind. Microbiol. Biotechnol. 32:639–665.

[163]De Souza, MP, Pilson-Smits EAH, Lytle CM, Hwang S, Tai J. Honma TSU, Yeh L, Terry N (1998). Rate-limiting steps in selenium assimilation and volatization by Indian Mustard. Plant Physiol. 117:1487–1494.

[164]Dehesh, K. (2004). Nucleic acid sequences encoding beta-ketoacyl-ACP synthetase and uses thereof. United States Patent Application 20040132189. Available from: http://www.patentstorms/patents/6706950.html (verified 23 January 2006).

[165]Dietz, A. C. and J. L. Schnoor (2001). " Advances in Phytoremediation." Env. Health Perspectives 109 (51): 163–168.

[166]El-Sayed AA (2006), El-Minia University, El-Minia, Egypt. Available at: Scolar.lib.vt.edu/thesis/.../Amr_ElSayed_Dissertation-bodytext.pdf

[167]EPA (2001). Treatment technologies for Site Cleanup: Annual Status Report, EPA.

[168]Fitzpatrick, K. and R. Scarth.1998. Improving the health and nutritional value of seed oils. PBI Bulletin. NRC-CRC. January:15–19.

[169]Fulekar, M. H., Anamika Singh and Anwesha M. Bhaduri (2009). Genetic engineering strategies for enhancing phytoremediation of heavy metals. African Journal of Biotechnology Vol. 8 (4), pp. 529–535. Available from: http://.www.academicjournals.org/AJB. ISSN 1684–5315 ©2009 Academic journals.

[170]Gilissen, J.W.J Luud, Suzanne T.H.P. Bolhar, Andre C. Knulst, Laurian Zuidmeer, Ronald van Ree, Z. S. Gao, and W. Eric van de Weg (2004). Production of hypoallergenic plant foods by selection, breeding and genetic modification. NJAS Wageningen Journal of Life sciences, pp. 1–11.

[171]Gilissen, LJ., Bolhaar ST., Matos, CI., et al., (2005). Silencing the major apple allergen Mald 1 by using the RNA interference approach. Journal of Allergy and Clinical Imm. 115 (2). 354–369.

[172]Hanson, R. I. (1991). "Evapotranspiration and Droughts, in Paulson." National water Summary 1988–89-Hydrologic Events and Floods and droughts: 99–104.

[173]Heilmann, W (2001) Technology II: Decaffeination of Coffee. In: Clarke RJ and Vitzthum OG (eds) Coffee Recent Developments, Blackwee Sciences, Oxford, UK pp.108-124.

[174]Henry, J. R. (2000). An overview of the phytoremediation of lead and mercury, EPA, Office of Solid Waste and Emergency Response, Technol. Innovation Office.

[175]Herman, E.M., R.M. Helm, R. Jung, and A.J. Kinney. 2003. Genetic modification removes an immunodominant allergen from soybean. Plant Physiol. 132:36–43.

[176]Hinchman, R. R. (1996). Phytoremediation: Using Green Plants to Clean Up Contaminated Soil, Groundwater And Wastewater. International Topical Meeting on Nuclear and harzadous Waste Management, Seatle WA.

[177]Hollander, D. (1997). Environmental effects of reproductive health: the endocrine disruption perceptive. Family planning perspectives, special reports 129 (2), 83–89.

[178]JAPAN: Biotech rice to prevent hypertension. Available from:http://bites.ksu.edu/news/758/09/05/22/japan-biotech-rice-prevent-hypertension. Publication: Crop Biotech Update - 22.may.09

[179]Jauhar, P. Prem (2006). Modern Biotechnology as an Integral Supplement to Conventional Plant Breeding: The Prospects and Challenges. Published in Crop Sci 46:1841-1859 (2006)© 2006 Crop Science Society of America 677 S. Segoe Rd., Madison, WI 53711 USA.

[180]Kato, M., K. Mizuno, A. Crozier, T. Fujimura, and H. Ashihara. 2000. Plant biotechnology: Caffeine synthase gene from tea leaves. Nature 406:956–957.

[181]Kramer, U, Chardonnens AN (2001). The use of transgenic plants in the bioremediation of soils contaminated with trace elements. Appl. Microbiol. Biotechnol. 55:661–672.

[182]Lay, D. J. (1999). "Phytoremediation of Trichloroethylene (TCE)." Retrieved 9/17/2009, from: http://www.horticulture.coafes.umn.edu/vd/h5015/99fpapers/lay.htm.

[21]Leighton Jones, publication manager: BMJ. 1999 February 27; 318 (7183): 581–584. Copyright © 1999, British Medical Journal - Science, medicine, and the future - Genetically modified foods.

[183]Lin CF, Lo SS, Lin HY, Lee YC (1998). Stabilization of cadmium contaminated soils using synthesized zeolite. J. Harvard. Mat. 60:217–226.

[184]Mariani, C., De Beuckeleer, M., Truettner, J., Leemans, J. & Goldberg, R. B. (1990). Introction of male sterility in plants by a chimeric ribonuclease gene. Nature 347:737–741.

[185]Miller, R. R. (1996). Phytoremediation: Technol. Overview Report. GWRTAC, Ground Water Remediation Technologies Analysis Center: 26.

[186]Murphy J. Denis (1999). Production of novel oil in plants: doi:10.1016/S0958-1669(99)80031-7.

[187]Ohar, K. Kokado Y. Yamamoto H., Sato F., Yazaki K. (2004). Engineering of ubiquinone biosynthesis using the yeast coq2 gene confers oxidative stress tolerance in transgenic tobacco. Plant J. 40:734–743.

[188]Ogita, S., Uefuji H, Yamaguchi Y, Koizumi N, Sano H (2003) Producing decaffeinated coffee plants. Nature 423:823.

[189]Ogita, S., Uefuji H, Morimoto M, Sano H (2004) Application of RNAi to confirm theobromine as the major intermediate for caffeine biosynthesis in coffee plants with potential for construction of decaffeinated varieties. Plant Mol. Biol. 54:931-941.

[190]Pilcher, H.R. (2003). GM decaf coffee grown on trees. Nature News Service/Macmillan Magazines Ltd. 19 June 2003.

[191]Rosser, S.J., C.E. French, and N.C. Bruce. 2001. Engineering plants for the phytodetoxification of explosives. *In* Vitro Cell. Dev. Biol. 37:330–333.

[192]Scarth Rachel and Jihong Tang (2006). Modification of Brassica Oil Using Conventional and transgenic Approaches. Published online April, 2006: http://www. crop.scjournals.org/cgi/reprint/46/3/1225.

[193]Silvarolla M.B., P. Mazzafera and L.C. Fazuoli (2004). A naturally decaffeinated arabica coffee. Nature 429:826.

[194]Simopoulos A. P. (2003). Importance of the ratio of omega-6/to omega-3 essential fatty acids: evolutionary aspects. World's Rev. of Nutrition and Diabetics. Ed. A. P. Simopoulos, C Leland LG vol. 92.

[195]Simple and essential nutritional requirement to control hypertension - the silent killer non drug treatment of hypertension . Indian J Med Sci 1999;53:444-56.

[196]Simple and essential nutritional requirement to control hypertension – the silent killer non drug treatment of hypertension. Indian J med Sci [serial online] 1999 [cited 2009 Jun 27]: 53:444–56. Available from: http://www.indianjmedsci.org/text.asp?1999/53/10/444/12230

[197]Schnoor, J. R. (1997). Phytoremediation: Evaluation report GWRTAC Pittsburgh, Pennsylvania, Ground-Water Remediation Technologies, Analysis Center.

[198]Schnoor, J. R. (2002). Phytoremediation of Soil and Groundwater, GWRTAC, Ground-Water Remediation Technologies, Analysis Center: 26.

[199]Udo Conrad 2005. Polymers from plants to develop bio-degradable plastics. Research Focus: doi:10.1016/j.tplants.2005.09.003.

[200]Vaijayanti Gupta (2006). From plants to plastics. SCIENCE/ ENVIRONMENT: http://www.indiatogether.org/2006/jun/env-plastics.htm

[201]Van Beilen, Jan B., Yves Poirier (May 2008). "Harnessing plant biomass for biofuels and biomaterials:Production of re-newable polymers from crop plants". *The Plant Journal* 54 (4): 684–701. doi:10.1111/j.1365-313X.2008.03431.x

[202]Van Ree R. (1997). The oral allergy syndrome. In: Amin, S., S., Lahti, A. Maibach, H. I. ed. Contact Urticania Syndrome. CRC Press, Boca raton, 289–299.

[203]Vogel, J.M. (2004). Tunnel vision: The regulation of endocrine disruptors. Policy Sciences; 37, 277–303.

[204]Wang Tao, YE Liang and Song Yanru (1999). Progress of PHA production in transgenic plants. Chinese Science Bulletin vol. 44. No. 19.

[205]Wong, K. 2001. Transgenic tobacco detoxifies TNT. Sci. Am. Dec 3, 2001.

CHAPTER VIII

Transgenic Plants Tolerant To Environmental Stress

Abiotic stresses, including drought and salinity, are estimated to cause yield losses worldwide of more than 50%. [210] Transgenic approaches offer an option to enhance drought and salt tolerance. [206, 207, 214] In order for us to understand the biotechnology that has been used to improve environmental stress in plants; we need to help answer the following question: What can plants do when they encounter or are exposed to environmental conditions that are less than optimal? There are two options: [208, 209, 211, 216, 225]

1. They can escape. This is not physical escape by getting away from the hostile environment, but rather a strategy to avoid stress by rapidly completing the life cycle of the plant. As the environment changes, the plant responds by initiating early flowering and seed production. By the time the stress is severe, the plant has reproduced and then dies or senesces.

2. They can adapt by making changes that allow the plant to survive and continue to grow even under sub-optimal conditions.

The process of adaptive responses of plants to environmental change is a complex field. Plants vary in the ways they adapt to the same environmental change. A common response is the activation of some protective mechanism(s). These protective mechanisms can take a variety of forms including the synthesis of proteins that have protective properties and the activation of metabolic pathways to synthesize protective compounds. Manipulating and altering these responses has been the focus of several research and development programs, with the goal of improving the stress tolerance of plants.

Listed below are some examples of biotechnological measures that have been used to alter the plants' response to environmental stress:

(a). Manipulating the synthesis of compatible solutes to protect against salinity and other stresses.

(b.) Constitutive activation of cold responses to protect against freezing injury.

(c.) Production of citrate to chelate aluminum and prevent aluminum toxicity.

(d.) Activation of enzymatic mechanisms for general protection against oxidative damage

Reports show that there are no commercial products on the market that result from these manipulations, and it may take a number of years before these types of products become available.

TRANSGENIC PLANTS SHOWING IMPROVED WATER DEFICIT AND OSMOTIC STRESS

While drought and salinity appear to be quite different environmental conditions; the two are related in the sense that they cause similar physiological problems for plants. In both cases, these stresses make it more difficult to keep water inside plant cells. Water will move out of the cytoplasm of the cell when the salt concentration is higher outside the cell than inside. How do plants deal with stress imposed by a lack of water? Water deficit is frequently the result of drought conditions. Another condition that also makes it difficult for plants to acquire water is high concentrations of salt in the soil, referred to as salinity stress. This can be caused by irrigation. For example, reports show that in Egypt, half of irrigated croplands suffer from salinization. [217, 221, 223]

It has been established that when plants are faced with water deficit and/or salinity stress; one response is that cells synthesize certain compounds referred to as compatible solutes which will allow water to be retained inside the cell.

These compounds must not be toxic and should allow normal metabolic processes to continue inside the cell. Examples of these compounds include mannitol (a sugar alcohol), glycinebetaine (abbreviated as GB) and the amino acid proline. Yeast accumulates glycerol when it experiences similar environmental conditions. Plants differ in the solutes they produce in response to water deficit. Some accumulate mannitol, others accumulate glycinebetaine. A biotechnological strategy that has shown some promise is to provide transgenic plants with the capacity to produce an osmotic solute that they don't normally make, or to increase their capacity to accumulate a solute that they normally produce. Such plants are under development using genes from bacteria. For example, researchers demonstrated that wheat engineered with the *mtlD* gene from *Escherichia coli* had improved tolerance to water stress and salinity. [179, 206, 225, 227]

Mannitol-Producing Plants

A bacterial gene called mtlD, from E. coli, catalyzes the conversion of fructose-6-phosphate to mannitol-6-phosphate. Plants contain the substrate for this reaction, and they also contain an enzyme (a phosphatase) that converts mannitol-6-phosphate to mannitol. So, simply providing the mtlD gene should allow plants to make mannitol. A modification of this approach would be to target the protein encoded by this bacterial mtlD gene to the chloroplast, an organelle that is particularly sensitive to salt damage. Transgenic plants with this modification are reported to have some increased salt tolerance. Reports show that Dekalb, the seed company that was acquired by Monsanto in 1998, has been conducting field trials of transgenic corn engineered to produce mannitol as a protection against drought stress and the accompanying water deficit. [206]

GENE MANIPULATIONS TOWARDS INCREASING CROP YIELDS

Plants that are capable of withstanding climate changes are likely to produce greater yields during inclement weather conditions. By modifying ^{heat-shock factors} (factors that allow misfolded proteins to resume their proper shape following damage by high temperatures during summer), genetists have recently managed to create heat tolerant cress plants. The GM plants are more robust and generally tolerate salty soils, hot and cold temperatures, and drought. Because crops have similar heat-shock factors, similar alterations could eventually lead to novel crops capable of producing greater yields during unusually hot, dry or cold seasons. Reports show that scientists at Agricultural and Agri-Food Canada have identified two genes in alfalfa that are directly linked to cold tolerance. The loss of alfalfa from exposure to cold weather is estimated at $10 million a year in Quebec alone and this application of this biotechnology may help to prevent such losses. [220, 224]

Another way of increasing crop yields is to trick the plants into thinking that autumn has not arrived. In autumn, plants usually pull useful proteins out of their leaves for storage in seeds or in their trunk or stem (this causes the leaves to change color), before allowing the leaves to drop off. This process is initiated when plant genes producing specific enzymes are "turned on." To prevent these enzymes from shutting down the plant's production capability, scientists at the university of Wisconsin-Madison added a gene that gets turned on at the same time as the "shut down" genes. This new gene produces an enzyme that makes a hormone called cytokinin. This hormone encourages leaves to stay green. The presence of cytokinin nullifies the "shut down" message, and keeps the leaves green well into autumn, extending the plant's growing season. It was also found that these modified plants stayed greener and fresher for a longer time after having been cut. This technique could eventually be applied to crops, where the extra weeks of photosynthesis could boost yields of grain and flowers, and could extend the shelf life of leafy vegetables like lettuce and cabbage. [220]

TRANSGENIC PLANTS TOLERANT TO COLD

Some plants are able to survive low temperatures whereas others are not. While it is not known precisely what is required for cold tolerance, it is known that plants respond to low temperatures in a number of ways. These responses include synthesizing a large number of proteins that are not expressed under normal conditions. The cold-induced proteins have been shown to function in helping the plant survive low temperature conditions.

In a bid to create plants expressing the genes encoding cold-induced proteins; scientists have identified the activator or control mechanism that activates all the genes that are expressed when it gets cold. It was established that the activator is a protein that switches on transcription of cold-induced genes, referred to as a transcription activator. At low temperatures; the transcription activator is induced; and cold-induced genes are expressed. With this knowledge in hand, it has been possible to engineer plants that constitutively express this suite of cold-induced proteins; using the following procedure: [215, 218, 225, 226]

The gene encoding the activator protein has been cloned. Normally this gene is itself activated by low temperatures; but the promoter of this gene has been altered so that it is no longer induced by low temperatures but is constitutively (all the time) expressed. Transgenic plants with this modified activator now express constitutively the suite of genes that are normally induced by low temperatures. Experimentally, these transgenic plants were able to survive when they were moved straight from normal conditions to extremely low temperatures up to -50° C, unlike normal plants. These results demonstrate that this approach of modifying the regulator or controller of a set of genes provides an excellent method to manipulate gene expression and perhaps alter and improve the properties of plants.

It certainly is useful to improve the cold tolerance of crop plants. It is a common observation that in the spring, as temperatures begin to rise many plants lose their ability to withstand low temperatures and become susceptible to late frosts. A good example is the production of peaches in Indiana. Growers cannot guarantee a good crop because large amounts of the crop are damaged from late frosts. If these and other plants could be genetically protected from this cold injury, it might be able to expand the area of production for these crops and improve the reliability of crop production in temperate regions.

TRANSGENIC PLANTS TOLERANT TO ALUMINUM TOXICITY

Aluminum toxicity and poor phosphorus availability are factors that limit plant growth on many agricultural soils. Aluminum is the most abundant metal ion in the rhizosphere and it is not required by plants, but is very toxic to most plants if it is available in the soil. The problem of Aluminum toxicity is not felt much in places where the soil is not acid. But where the soil has a low pH (acid conditions) Aluminum is soluble in the soil solution and can severely inhibit plant growth. [213, 219]

Aluminum is highly toxic to the growing region of root tips but has little effect on other parts of the root. Over the last few years it has been shown that perhaps the most important mechanism of Aluminum tolerance is one that prevents it from being taken up by the root tip. The logic behind the functioning of this mechanism is that if the root tip is protected, then the rest of the root system is able to continue functioning. Tolerant plants are able to reduce Aluminum uptake by secreting organic acids into the soil around the root tip. These organic acids (malate and citrate) are able to chelate (bind up, sequester) the Aluminum that is in the soil solution right around the root tip. If the Aluminum is bound to one of these organic acids, it cannot enter the plant root and cause damage. [212, 219, 222]

Reports from previous studies showed that the expression of a bacterial gene *Pseudomonas aeruginosa* citrate synthatase gene in a plant - tobacco (*Nicotiana tabacum*; CSb lines) resulted in improved Aluminum tolerance. The citrate produced in the roots is able to get out of the root and into the rhizosphere around the root where it chelates the Aluminum. These transgenic plants have been shown to be more tolerant of Aluminum under acid conditions. Researchers suggest that if this approach can be developed successfully in crop plants, it will provide a method to improve Aluminum tolerance in many crops and perhaps to improve the agricultural productivity of crops grown under acid soil conditions. This technology may find widespread application in the tropics where acidic soils are widespread. [212, 219, 222]

LITERATURE CITED

[206]Abebe, T., A.C. Guenzi, B. Martin, and J.C. Cushman. (2003). Tolerance of mannitol-accumulating transgenic wheat to water stress and salinity. Plant Physiol. 131:1748–1755.

[207]Apse, MP., and Blumwald, 2002. Engineering salt tolerance in plants. Curr. Opin. Biotechnol. 13:146–150.

[208]Bacon, MA (2004). Water Use efficiency in plant Biology, Blackwell.

[209]Boyer JS (1982). Plant productivity and environment. Science 218:443–448.

[210]Bray, E.A., J. Bailey-Serres, and E. Weretilnyk (2000). Responses to abiotic stresses. p. 1158–1249. *In* W. Gruissem et al. (ed.) Biochemistry and molecular biology of plants. Am. Soc. of Plant Physiol., Rockville, MD.

[211]Cushman JC, Bohnert HJ (2000). Genomic approaches to plant stress tolerance, Curr Opin plant Biol. 3:117–124.

[212]De la Fuente, J.M. V., Ramírez-Rodríguez, J.L. Cabrera-Ponce, L. Herrera-Estrella [1997] Science 276: 1566–1568.

[213]Delhaize E. CSIRO Division of Plant industry, GPO Box. 1600. Canberra ACT 2601, Australia. Extracted from: http://www.plantstress.com/Articles/toxicity_m/Tolerance.htm

[214]Flowers, T.J. 2004. Improving crop salt tolerance. J. Exp. Bot. 55:307–319.

[215]Guy CL. (1990). Cold acclimation and freezing stress tolerance: Role of protein metabolism. Annu Rev Plant Physiol Plant Mol Biol. 1990;41:187–223.

[216]Grime JP (2001). Plant Strategies, Vegetation Processes, and Ecosystem Properties. John Wiley & Sons.

[217]Jacob W. Kijne (2005) - Abiotic stress and water scarcity: Identifying and resolving conflicts from plant level to global level. Copyright © 2005 Elsevier B.V. All rights reserved.

[179]Jauhar P. Prem (2006). Modern Biotechnology as an Integral Supplement to Conventional Plant Breeding: The Prospects and Challenges. Published in Crop Sci 46:1841-1859 (2006)© 2006 Crop Science Society of America 677 S. Segoe Rd., Madison, WI 53711 USA.

[218]Kasuga M., Liu Q., Miura S., Yamaguchi-Shinozaki, K. (1999). Improving plant drought, salt and freezing tolerance by gene transfer of a single stress-inducible transcription factor. Nat. Biotechnol. 17:287–291.

[219]Ma JF, Ryan PR, and Delhaize E (2001) Aluminium tolerance in plants and the complexing role of organic acids. Trends Plant Sci 6: 273–278.

[220]Mittler, R. (2005). Abiotic Stress, the field environment and stress combination. Trends Plant Sc. 11, 15–19.

[221]Rebaut J-M & D. Poland (eds) 2000. Molecular Approach for the Genetic Improvement of cereals for Stable Production in Water-limited Environments. A Strategic Planning workshop held at CIMMYT, El Batan, Mexico, 21–25 June 1999. Mexico D. F.: CIMMYT.

[222]Ryan PR, Delhaize E, and Jones DL (2001). Function and mechanism of organic anion exudation from plant roots. Annu Rev Plant Physiol Plant Mol Biol 52: 527–560.

[223]Stephen Leahy (2005). "Environment: Millions Flee Floods, Desertification", I.P.S., Brooklin, Canada, 10/12/05.

[224]Varshney, R. K. and R. Tuberoza (eds) (2007). Genomics-Assisted Crop Improvement, Vol. 1: Genomics Approaches and Platforms, 1–12. 10.10071978-1-4020-6295-7-1.

[225]Vinocur, B., and A. Altman. 2005. Recent advances in engineering plant tolerance to abiotic stress: Achievements and limitations. Biotechnology (NY) 16:1–10.

[226]Wanner A. Leslie and Olavi Juntilla (1999). Cold-induced Freezing Tolerance in *Arabidopsis*. Plant Physiol. 1999 June, 120 (2): 391–400.

[227]WO 1992019731 19921112: Transgenic plants With altered Polyol Content.

CHAPTER IX

Gene Cloning

Methodologies regarding gene cloning are well documented: Ref.: [152, 179, 228, 229, 233, 234, 235, 236, 238, 239, 241, 242]

Gene cloning encompasses a technology whereby genes are transferred into a variety of different organisms, from microbes to plants and mammals. This technology forms the basis for the production of transgenic animals and plants. A transgenic organism (animal or plant) is an organism that has been successfully transformed and contains a new gene or genes which are stably inherited by its progeny. As previously noted, these are also known as transformants or genetically modified organisms (GMOs).

The five basic events in a gene cloning experiment are as follows:

1. A fragment of DNA containing the gene to be cloned is inserted into a second (usually circular) DNA molecule, called a cloning vector, to produce a recombinant DNA molecule.

2. The recombinant DNA molecule is then introduced into a host cell (often the bacterium, *Escherichia coli*) by transformation.

3. Within the host cell, the vector directs multiplication of the recombinant DNA molecule, producing a number of identical copies.

4. When the host cell divides, copies of the recombinant DNA molecule are passed onto the progeny and further vector replication takes place.

5. A large number of cell divisions give rise to a clone, a colony of cells each containing multiple copies of the recombinant DNA molecule.

ENZYMES USED IN GENE CLONING

The first step in gene cloning or the construction of recombinant DNA molecules involves cutting DNA molecules at specific sites and joining them together again in a controlled manner. These manipulative techniques make use of purified enzymes that, in the cell, participate in processes such as DNA replication and protection. The two main types of DNA manipulative enzymes used in gene cloning are restriction endonucleases and DNA ligases. Restriction endonucleases are enzymes that cut DNA.

During construction of a recombinant DNA molecule, in order to add other pieces of DNA to a plasmid vector, the vector must contain a site where the circular DNA molecule can be cut with a restriction enzyme to open the molecule. Other DNA fragments can then be inserted into the vector using DNA ligase enzyme which joins the DNA molecules and creates a new circular plasmid DNA.

The circular vector must be cut at a single point into which the fragment to be cloned can be inserted. Not only must each vector be cut just once, but all the vector molecules must be cut at precisely the same position. Therefore a very special type of nuclease is needed to accomplish this. The relevant enzymes are called Type II restriction endonucleases. Each recognizes a specific nucleotide sequence [the DNA molecule is made up of the following nucleotide bases: Adenine (A), Guanine (G), Thymine (T) and Cytosine (C)], and cuts a DNA molecule at this sequence and nowhere else. For example, the restriction endonuclease called EcoRI (isolated from *E. coli*) cuts DNA only at the hexanucleotide GAATTC. In contrast, a second enzyme from the same bacterium, called EcoRV, cuts at a different hexanucleotide, in this case GATATC.

Many restriction endonucleases recognize hexanucleotide target sites, but others cut at four-, five- or even eight-nucleotide sequences. There are also examples of restriction endonucleases with degenerate recognition sequences, meaning that that they cut DNA at any of a family of related sites. Hinfl, for example, recognizes GANTC, so it cuts at GAATC, GATTC, GAGTC and GACTC. Thus the exact nature of the cut produced by the restriction enzyme is very important in the construction of recombinant DNA molecules. Many restriction endonucleases make a simple double stranded cut in the middle of the recognition sequence, resulting in blunt ends.

Other restriction enzymes cut DNA in a slightly different way. With these enzymes the two strands are not cut at exactly the same position in the two DNA strands. Instead the cleavage is staggered, usually by two or four nucleotides, so that the resulting DNA fragments have short single-stranded overhangs at each end. These are called sticky or cohesive ends because base pairing between them can stick the DNA molecules back again.

The final step in the construction of a recombinant DNA molecule is the joining together of the vector molecule and the DNA fragment to be cloned. This process is referred to as ligation, and the enzyme that catalyses the reaction is called DNA ligase. All living cells produce DNA ligases but the enzyme most frequently used in gene cloning is the one involved in replication of the bacterial virus T4 phage. Within the cell, the enzyme synthesizes phosphodiester bonds between adjacent DNA fragments during replication.

In the test tube, purified DNA ligases carry out exactly the same reaction and can join together broken DNA fragments in either of two ways, the first of which is by repairing the discontinuities in the base-paired structure formed between two sticky ends. This is relatively efficient because the ends of the molecules are held in place, at least transiently, by sticky-end base pairing. The second way involves joining together two blunt-ended molecules. This is a less efficient process because DNA ligase has to wait for chance associations to bring the ends together.

THE PROCESS OF TRANDFORMATION

These recombinant DNA molecules are normally formed in a test tube and are then transferred into a bacterium by a process known as transformation. Transformation can be simply defined as the transfer of DNA into a cell. The host or recipient is the organism that is going to take up the DNA and be transformed. In order for this process to succeed, the host must be competent for transformation, that is, have the ability to take up DNA.

A variety of treatments are used to make bacterial cells competent to take up DNA. One of these is treating the bacteria with ice cold calcium chloride or polyethyleneglycol. It is not fully understood what this treatment does to the bacteria, but it is assumed that the physical barriers to DNA getting into the bacteria, which are the cell wall and membrane, are made more permeable thereby allowing DNA to enter the cell.

It has been found that in almost all transformation systems, the frequency of transformation or rather the fraction of cells that are transformed and take up DNA is low. For example, in the bacterium *E. coli*, the highest transformation frequency that can be obtained is about 1%. In other words, at best only 1 bacterium out of every 100 will be transformed.

After transformation, it is important for the transformed bacteria to be identified so that they are distinguishable from the others. The DNA that the bacteria take up should carry a gene that gives the transformed bacteria a distinct property or phenotype that can distinguish them from the non-transformed bacteria. In transformation of bacteria, the DNA used for transformation carries a gene for resistance to an antibiotic, such as ampicillin or tetracycline. After mixing the competent bacterial cells and the DNA, the bacteria are plated on an agar medium that contains the antibiotic and allowed to grow. Only the bacteria that have been transformed with the DNA will be able to grow in the presence of the antibiotic, while untransformed bacteria (the vast majority) will not grow.

This antibiotic resistance gene is also known as a selectable marker, because it allows for direct selection of transformed bacteria. Selectable marker genes are essential for transformation of virtually all organisms because the frequency of transformation is very low. Transformation with vectors that carry a selectable marker allows the transformed organisms to be simply identified.

The purpose of transforming a bacterium or a yeast cell is to produce multiple copies of the gene of interest. The genes can then be moved into the host organism (plant or animal), with the aid of a vector, much in the same way that vectors carry infectious diseases between organisms. Several types of naturally occurring DNA molecules have been adapted to be used as cloning vectors: viruses and plasmids. Literally, hundreds of different cloning vectors are now available for use with different types of host cells.

As for microorganisms, a number of methods have been developed for transformation of plants. These methods utilize either biological or physical methods to deliver DNA into plant cells. For most eukaryotes, extra-chromosomal plasmid DNA molecules have either not been identified or not been developed as vectors for transformation. Therefore, most of the transformation methods used for higher eukaryotes rely on integration of the DNA that is transferred into a chromosome. When the DNA is integrated into the chromosome it will be reliably transferred through mitosis and meiosis, and stably inherited.

Transformation methods can be divided into two categories. One category uses purely physical methods to get DNA into an organism. The calcium chloride or other chemical treatments used to make cells competent fall into this category. The other category uses a biological agent such as a bacterium or virus, as the vector to transfer the DNA. Whichever method is used, the DNA must cross a number of physical barriers for transformation to be successful. For transformation of a plant, the overhead barriers to be overcome include the cell wall, plasma membrane, contents of the cytoplasm, nuclear membrane, and ultimately the integration into a chromosome.

All of the transformation methods for plants follow a general approach. Tissue samples are removed from the host plant which is to be transformed. These samples are known as explants and are taken from a variety of tissues. Alternatively, the source of plant material for transformation can be tissue culture, where plant cells are grown under aseptic conditions on an artificial agar medium. The plant material is then subjected to the transformation procedure. The goal is to transform as many cells as possible in the tissue. Many studies have shown that most transformation procedures are inefficient and only a small fraction of the target cells are actually transformed. Selection is applied to identify the rare cells that have been transformed. Only transformed cells will be able to grow under these selection conditions.

The composition of the growth medium, especially the levels of auxin and cytokinin hormones, is altered so that differentiated plants will regenerate from transformed cells. Seeds are then collected from the regenerated, transformed plants and the progeny are examined for inheritance of the transgene (the gene that has been transferred).

BIOLOGICAL METHODS USED IN PLANT TRANSFORMATION

Ref.: [232, 235, 239]

The first reliable method for plant transformation was based on a pathogen that attacks plants, especially grape and olive, and causes crown gall disease - the formation of galls at the crown of a plant, or the junction between root and shoot at the soil surface. The organism that causes this disease is *Agrobacterium tumefaciens*; loosely translated from the Latin as "soil-bacterium tumor-maker". The galls are produced at the site of infection and consist of a mass of undifferentiated cells, also known as tumors. *Agrobacterium* produces these tumors by transferring a piece of DNA from the bacterium to the plant. This is called the T-DNA for "transferred DNA".

The T-DNA carries a number of genes with promoters and other control sequences that are designed to function in plants rather than in bacteria. The genes on the T-DNA are then expressed in the plant. The T-DNA genes encode enzymes to make the plant hormones auxin and cytokinin. The concentration and balance between auxin and cytokinin must be controlled since higher levels of auxin promote root growth, while cytokinin promotes shoot development. The T-DNA genes result in very high levels of both hormones, and this leads to the proliferation of a tumor. The T-DNA in the bacterium is part of a large plasmid known as the tumor-inducing or Ti plasmid. Bacteria which do not have the Ti plasmid cannot produce galls and are therefore not pathogenic.

The T-DNA also carries genes encoding enzymes for the production of opines, unusual amino acid derivatives. The tumor produces opines and bacteria (living in the tumor outside the plant cells) use these compounds as their source of carbon and nitrogen. Some of the other genes carried by the Ti plasmid encode enzymes to utilize these opines in metabolism. By producing opines the tumor provides an ecological niche for the *Agrobacterium* to grow. *Agrobacterium* has developed this unique method to transfer DNA from the bacterium into plants. This has now been manipulated to develop systems for transformation of plants. The T-DNA region of the Ti plasmid, the portion that is transferred into plant cells, is physically defined by specific sequences of approximately 20 base pairs on either end of the T-DNA. Any DNA that lies between these border sequences will be transferred into the plant.

Several researchers have established that although the T-DNA contains genes for hormone synthesis and production of opines. These genes are not required for the transfer of DNA from the bacterium. That is to say, the bacterium is not aware of the sequence of DNA that is being transferred. This is important for two reasons. First, any piece of DNA can be placed between the border sequences and will be transferred form bacterium to plant. Secondly, the genes that result in altered synthesis of hormones, thereby leading to the production of tumors, can be removed. Therefore, it is possible to obtain transformed cells which have normal levels of hormones and appear to be quite normal.

While plant cells can be readily transformed by *Agrobacterium,* there must be some method to distinguish between transformed cells and those that are not transformed. The fraction of cells that are transformed is small. Without any selection, most of the regenerated plants would not be transformed and some other method would have to be used to identify transformed cells and plants. It is therefore necessary that the T-DNA that is transferred to plants must also carry a gene giving a selective advantage to transformed plant cells under specific conditions. This means a selectable marker must be included in the T-DNA. For plants, the first selectable marker gene was for kanamycin resistance. The gene for resistance was found in bacteria. This gene encodes a protein that adds a phosphate group to kanamycin, thereby inactivating the antibiotic.

Growth of plant cells has been shown to be inhibited by kanamycin. Therefore if the bacterial protein that inactivates kanamycin is expressed in transgenic plants, then the plants will be able to grow in the presence of kanamycin. In order to express the protein encoded by this bacterial gene in plants, the gene has to be modified first. These modifications include removing the promoter and terminator sequences which direct the transcription of this gene in bacteria. They are then replaced with a promoter from a plant gene which directs a high level of expression of the bacterial gene in all plant cells and a transcription terminator which also works in plant cells. For example, a kanamycin resistance gene for plant transformation can be made by first inserting the chimeric resistance gene into the T-DNA of the Ti plasmid so that it will be transferred from bacterium to plant. A selectable marker gene for kanamycin resistance would then be added to the T-DNA between the border sequences. The Ti plasmid can then be further manipulated so that the T-DNA genes that normally lead to production of tumors (i.e. the genes that direct synthesis of hormones) are removed.

Any other gene or genes can be included in the T-DNA and will be transferred from the bacterium into the plant. Using this modified Ti plasmid in *Agrobacterium tumefaciens*, transgenic plants can be produced by first removing the tissue explants from the plant, and incubating the wounded plant tissue with *Agrobacterium* carrying the modified Ti plasmid. The infected tissue is then placed in a medium containing kanamycin. The transformed cells divide and grow, and are then transferred to a special medium to allow shoots to develop. The shoots are placed on another medium to promote root development and produce small plantlets. The plants are then transferred from agar medium to soil. This method is very widely used for plant transformation.

It has been found that when bacteria or yeast are transformed with a plasmid, every cell that takes up the plasmid is identical. But this is not the case when plant cells are transformed with *Agrobacterium*. Each plant that is obtained from a separate transformation event should be regarded as unique with a distinct genotype. One major difference is that the T-DNA must integrate into a chromosome, rather than replicate extra-chromosomally. The site of integration appears to be random. Because the T-DNA is integrated at different sites in different cells, each transformed cell is now unique. There are two reasons why this random integration is important. First the T-DNA may insert into, and inactivate, a gene that is required for some other function or purpose. While this insertion may not be immediately obvious, it may result in the inactivation of perhaps a gene for resistance to a particular pathogen.

The site of integration can have a very large effect on the level at which the T-DNA genes are expressed, and the expression can easily differ by a factor of 100 between individual transgenic plants, even though they contain exactly the same piece of DNA. Secondly, the process of tissue culture that involves growing plant material in culture, exposing it to hormones, antibiotics, and other exogenous agents, and regenerating differentiated plants can be a successful mode of producing a variation. This is known as somaclonal variation; a phenomenon that is well documented, but not well understood.

Wheat, rice, and maize are stable foods, and there is great economic incentive to develop transgenic varieties. It used to be thought that *Agrobacterium tumefaciens* was not very efficient at infecting these monocots. Many of these important agronomic crops cannot be transformed using this method for two notable reasons. One is the inability of the bacterium to transfer DNA into these plants; this may not be so surprising since after all not all plant species are susceptible to infection by *Agrobacterium*. The other reason is a failure to regenerate plants after transformation. Again this is perhaps not surprising because not all plant species are as easy to regenerate plants from single cells as tobacco or carrot.

In the last two years however, there have been a number of reports showing that *Agrobacterium* can be used to transform rice and maize at high efficiency. Successful use of *Agrobacterium* for transformation of maize and rice depends on improving several facets of the procedure including the choice of tissue for infection (immature embryos at a specific stage of development); the vector and strain of *Agrobacterium* (with a super-active T-DNA transfer mechanism), and the conditions used for culture of embryos after transformation.

PHYSICAL METHODS USED IN TRANSFORMATION

Since *Agrobacterium* transformation is not successful with some plant species, a variety of physical methods have been developed to deliver DNA into plant cells. The first method is virtually identical in principle to the methods used to transform microbes. Plant tissues can be reduced to a collection of individual cells that lack cell walls. This is done by treating plant tissue with a collection of enzymes that break down the cell wall polymers. These cells without any wall are called protoplasts. One of the major barriers to getting DNA into plant cells (the cell wall) gets removed in the preparation of protoplasts. The protoplasts can then be transformed using physical techniques such as electroporation, microinjection, and bolistic particle delivery system.

(a.) Electroporation [242]

In electroporation, the protoplasts are mixed with the DNA (the genes to be introduced) and are then given an electric shock. In some way this high voltage makes temporary holes or pores in the plasma membrane which allows DNA to enter the cell. The DNA then somehow gets to the nucleus and is integrated into a chromosome. Electroporation is generally the most efficient treatment to get DNA inside plant protoplasts, but other chemical treatments, such as addition of polyethylene glycol (PEG) or calcium chloride, can also be used. After transformation, plants must be regenerated from transformed cells, using procedures quite similar to those used in *Agrobacterium*-mediated transformation.

(b.) Microinjection [231, 237]

Another physical method to get DNA into protoplasts is by microinjection, using very fine needles and syringes to directly inject a solution of DNA into the nucleus of a cell. One at a time, individual protoplasts are held in place by suction from a pipette, while a needle is inserted through the cytoplasm into the nucleus. A few nanoliters of DNA solution are injected into the nucleus, and then whole plants are regenerated from these injected, transformed cells. One advantage of this method is that the frequency of getting DNA into the protoplasts is very high. Essentially all the cells that are injected get the DNA. However, the number of cells that can be transformed is limited.

(c.) **Biolistic Particle Delivery System** [230, 234, 238, 240]

This method is also commonly referred to as microprojectile bombardment ("shotgun"). In this method the DNA is delivered into cells on microscopic particles of gold or tungsten. The DNA is mixed with the particles (1 to 10 micron diameter) on which it precipitates. This leads to the formation of particles coated with gold. The DNA-coated particles are placed on the end of a larger plastic bullet. The bullet is loaded into a gun barrel and the target – plant tissue that is to be transformed is positioned at the end of the barrel. Subsequently, the gun is fired, accelerating the bullet to the end of the barrel. A plate with a small hole in the end stops the plastic bullet. However the small, DNA-coated particles pass through the hole and strike the target.

A general observation is that some of the DNA-coated particles will pass through the cell wall and enter individual cells. In electron microscope pictures of plant tissue that has been bombarded, the metal particles can be seen distributed in the tissue. Some of the DNA will be released from the particle, to end up in the nucleus where it integrates into a chromosome, resulting in transformation. Transgenic plants can then be produced from these transformed cells in a variety of ways.

Since the microprojectile method has been found to damage to the plant tissue, alternative methods have been introduced. The most widely used is gas pressure, normally provided by compressed helium. The primary advantage of particle bombardment is that it has been shown to have an unlimited host range. Any cell can be physically penetrated by these particles, and thereby transformed. As mentioned earlier, this method is not limited to plants. It has also been applied to animals. It has been found that the ability to regenerate plants from single cells in culture is not limiting. For example, immature embryos which have been removed from a developing seed are in the process of developing into a seedling, and can continue on this path after they have been bombarded with these particles. Particle bombardment can be used to transform cells in organized morphogenic tissues, such as meristems or embryos. These have the advantage that they are already in the process of developing into differentiated tissues. They are therefore not dependent on tissue culture procedures to induce differentiation, organogenesis or embryogenesis.

LITERATURE CITED

[228]Altpeter, F., A. Varshney, O. Abderhalden, D. Douchkov, C. Sautter, J. Kimlehn, R. Dudler, and P. Schweizer. (2005). Stable expression of a defense-related gene in wheat epidermis under transcriptional control of a novel promoter confers pathogen resistance. Plant Mol. Biol. 52:271–283.

[229]Borlaug, N. E. (1997): Feeding a world of 10 billion people: the miracle ahead. Plant Tissue Culture and Biotechnology, Vol. 3, No.3:119–127.

[230]Cho, M.J., H. Yano, D. Okamoto, H.K. Kim, H.R. Jung, K. Newcomb, V.K. Le, H.S. Yoo, R. Langham, B.B. Buchanan, and P.G. Lemaux. (2004). Stable transformation of rice (*Oryza sativa* L.) via microprojectile bombardment of highly regenerative, green tissue derived from mature seed. Plant Cell Rep. 22:483–489.

[231]Dunwell, J.M. (2000). Transgenic approaches to crop improvement. J. Exp. Bot. 51:487–496.

[232]Funatsuki, H., M. Kuroda, P.A. Lazzeri, E. Muller, H. Lorz, and I. Kishinami. (1995). Fertile transgenic barley generated by direct transfer to protoplasts. Theor. Appl. Genet. 91:707–712.

[179]Jauhar P. Prem (2006). Modern Biotechnology as an Integral Supplement to conventional Plant Breeding: The Prospects and Challenges. Crop Sci 46:1841-1859 (2006) © 2006 Crop Science Society of America.

[233]Kammermayor K. and Clark, V. L. (1989) - Genetic Engineering Fundamentals, An introduction to Principles & Applications, Marcel Decker Inc., 1989.

[234]Klein, T.M., and T.J. Jones (1999). Methods of genetic transformation: The gene gun. p. 21–42. *In* I.K. Vasil (ed.) Molecular improvement of cereal crops. Kluwer Academic, London.

[235]Komari, T., and T. Kubo (1999). Methods of genetic transformation: *Agrobacterium tumefaciens*. p. 43–82. I.K. Vasil (ed.) Molecular improvement of cereal crops. Kluwer Academic, London.

[236]Naisbitt John and Patricia (1990). Aburdene. Megatrends 2000. Pan Books.

[237]Neuhaus, G., Spangenburg, G. (1990). Plant transformation by microinjection techniques. Physiol. Plant 79:213–217.

[238]Patnaik, D., and P. Khurana. (2003). Genetic transformation of Indian bread (*T. aestivum*) and pasta (*T. durum*) wheat by particle bombardment of mature embryo-derived calli. BMC Plant Biol. 3:5.

[239]Ramesh, S., D. Nagadhara, V.D. Reddy, and K.V. Rao. (2004). Production of transgenic indica rice resistant to yellow stem borer and sap-sucking insects, using superbinary vectors of *Agrobacterium tumefaciens*. Plant Sci. 166:1077–1085.

[240]Sagi L, Panis B, Remy S, Schoofs H, De Smet K, Swennen R, Bruno PAC (1995). Genetic transformation of banana and plantain (Musa spp.) via particle bombardment. Biotechnol 13:481–485.

[241]Sambrook J, Fritsch EF, Maniatis T (1989). Molecular cloning: a laboratory manual, 2nd edn. CSHL Press, Cold Spring Harbor.

[242]Shillito, R. (1999). Methods of genetic transformation: Electroporation and polyethylene glycol treatment. p.

9–20. *In* I.K. Vasil (ed.) Molecular improvement of cereal crops. Kluwer Academic, London.

[152]Snow AA, Andow DA, Gepts P, Hallermann EM, Power A, et al., (2004). Genetically engineered organisms and the environment: current status and recommendations: http://www.esa.org/pao/esaPosition/Papers/geo_position.htm - 2009.

CHAPTER X

CURRENT STATUS ON THE ADOPTION OF GMOs

Genetically modified foods have certainly arrived. Scientists want to exploit this new technology, which they also claim to be necessary to help feed the world's burgeoning population. [255, 256, 267] Continued increases in crop yields are required to feed the world in the 21st century, [248, 258, 264; 254] given the continuing decline of areas suitable for grain production as a result of urbanization and industrialization.

While the worldwide demand for grain has been increasing along with an expanding population and a rising income level, the production of primary grains has not been sufficient to meet the demand, and the world stocks-to-use ratio, that is, the grain supply measured against grain demand, has been declining. In parallel, a shortage of water resources, continued global warming, and many other factors of instability have been threatening grain production. In addition, the increasing demand for biofuel may have a huge impact on the supply and demand for food in the world. [254, 258] Thus food security, especially in developing countries remains a challenge. [259] This challenge is made worse by the adverse effect of climate change in those countries. [259, 265] Climate continues to have major impact on crop productivity all over the world. For example, reports show that in China, rising temperature over the past two decades has led to a 2.4% decline in wheat yields. Therefore, adoption of crop varieties that are drought or cold tolerant can lead to increased crop yields. [259, 264]

There is a general assumption that transgenic plants that are now in use in many countries have passed through strict laboratory and field tests to show that they are safe for humans and not harmful to the environment. The GM foods that are already on supermarket shelves have been subjected to rigorous safety assessment, as will the many more currently under development. Reports show that the United Kingdom has led the world in developing systems for assessing the safety of genetically modified foods. These systems are based on rational scientific evaluation by leading experts, within the limits of current knowledge. Within the European Union genetically modified foods are now regulated on a union-wide basis – the European regulation on novel foods and novel food ingredients. [263] Apart from a thorough safety evaluation, Europe now requires mandatory labeling of GM containing foods. However, as of now, no labeling is required for animal feeds. [251, 252, 263]

In the United States, genetically modified plants/foods/medicines are evaluated, prior to their commercialization, by three regulatory agencies: The US Department of Agriculture (USDA); Center for food and Applied Nutrition (CFSAN), US Food and Drug Administration (FDA) and US Environmental Protection Agency (EPA). The EPA evaluates GM plants for environmental safety, the USDA evaluates whether the plant is safe to grow, and the FDA and CFSAN evaluates whether the plant is safe to eat. [253]

In the Philippines, research on genetically modified plants is regulated by the National Committee on Biosafety of the Philippines (NCBP). Strict guidelines ensure the safety of the public and the environment is guaranteed before granting permission for research in this field. Arguably, the Philippines has one of the most stringent biosafety guidelines in the world. [252]

There are two tests for the determination of the presence of GM ingredients and the measurement of the level of materials. These are: (a) Polymerase chain reaction (PCR) test for DNA; [243, 251, 258] and (b) Enzyme-linked immune-sorbent assay (ELISA) for protein. [243, 251] The PCR test is the most commonly used method to determine the presence of novel DNA in GM food. Sensitivity of PCR test decreases with further processing of the food product. However, for most products, the detection range is between: 0.1% to 1%. Testing cost per sample ranges from US$ 200–US$ 600, and it takes five to fourteen days to produce the result. The way the PCR works is to generate billions of copies of a single DNA molecule in a matter of hours. Through biochemical processes, a sample of the DNA is scanned to locate and label target sequences of DNA which are amplified billions of times. The amplification allows detection of a specific sequence and quantification of the proportion of DNA molecules in the sample. [243, 251, 258]

The ELISA test is designed to detect the presence of a novel protein in a GM food. Similar to PCR, conducting ELISA test requires trained personnel and specialized equipment. Testing cost per sample is relatively cheap, and ranges from US$ 75–US$ 100, and it takes two to four days to produce the result. The way that ELISA works is to use antibodies specific for the protein of interest. ELISA uses one antibody to bind the specific protein, a second antibody to amplify detection, and a third antibody conjugated to an enzyme whose product generates a color that can be easily visualized and quantified. [243, 251]

A general observation is that, the detection of GM foods in a given commodity is expensive and presently cannot be done in many developing countries. Thus, these countries have no legislation mechanisms restricting the release of GMOs to their markets. This may promote "dumping" of GM foods into such countries that rely a lot on food donations. In any case, with or without legislation, market forces will still prevail. If the GM seeds provide a better yield, unscrupulous farmers will go 'out of their way' to buy them. In cases where genetic modification has introduced substantial changes to foodstuffs, consumers are supposed to be informed through statutory (and additional voluntary labeling). However, analysts have found that it is not easy to label GM foods on supermarket shelves. One author is quick to note that, a general label such as, "This product might contain plant material which may have been subject to genetic modification" is not very helpful, as such labeling tends to "demonize" GM foods in the mind of the public. [251, 261]

Mandatory food labeling is thus a complex issue. For example, in the United States, the FDA's current position on food labeling is governed by the Food, Drug and Cosmetic Act, which is only concerned with food additives, and not whole foods or food products that are considered GRAS (Generally Recognized As Safe). It is further noted that the FDA contends that GM foods are substantially equivalent to non-GM foods, and therefore not subject to more stringent labeling. [251 252, 267]

There is no universally accepted agreement regarding GM labeling policies in the international community. Countries such as US and Canada choose voluntary labeling policy, while others, such as those of the European Union, Australia, New Zealand and Japan, adopt a mandatory labeling policy. [251] In a study conducted by Zepeda, [670] the food labeling policies were found to be correlated with the economic interest in GM crops of each jurisdiction. The contrasting positions of countries regarding labeling requirements seem to correspond closely to the production of GM crops for commercial sale. [251]

According to a study conducted by Buttel, [247] Professor of Rural Sociology and Environmental Studies at the University of Wisconsin-Madison, the US, together with Canada and Argentina, grow roughly 98% of all GM crops, and these three countries do not require mandatory labeling of GM food. With a significant economic stake in GM food production and sales, the US adopts a reactive labeling policy. The labeling of GM food is voluntary, because GM food is presumed to be GRAS. Labeling is only required when the food is significantly different from its conventional counterpart in characteristics, such as composition, nutrition, allergenicity or toxicology. [251]

In contrast, Australia grows a very small amount of GM crops, [247] while Japan does not grow any GM crops. [258, 261] Australia and Japan place great emphasis on the pre-market assessment of new GM varieties, GM products are examined and tested by enforcement agencies before they are allowed to be distributed in the market. Reports show that the relevant authorities in these two countries have taken the lead in implementing mandatory labeling of GM food. In Australia, labeling of GM food is mandatory if novel DNA and/or novel protein is present in the final food product by more than 1%. [247, 251] In Japan, labeling of GM food is mandatory if the food product contains any of the designated GM ingredients that is one of its top three ingredients and accounts for 5% or more of the total weight of the food product. [260] As of Oct. 2007, seven agricultural products, i.e., soy, maize, potatoes, canola, cotton, alfalfa, and sugar beets, and their processed foods have been designated as mandatory labeling items. [258] For food products containing GM ingredients which are not approved, it is illegal to either sell or import them regardless of the content percentage. [251, 258, 261]

In spite of the obvious discrepancies in the existing policies governing adoption of GM products, GM crops have been widely adopted in many developed countries including the United States. Developing countries like the Philippines, Brazil, India and China have also embraced the importance of using biotechnology to make themselves self-sufficient in food grains, including maize, rice and wheat. Reports show that in 2007, USA, Argentina, Brazil, Canada, India and China continued to be the principal adopters of biotech crops globally. [255] As of 2003, the global acreage of genetically modified maize was quite extensive, covering more than 15 million hectares. More than 70 biotechnology plant varieties have been commercialized in the USA. [262] In addition to this, the multi-million dollar losses from insect pests suffered by cotton farmers have been reduced by the use of Bt cotton. [256, 268] Many countries are now actively growing Bt crops that produce insecticidal toxins. Pest-resistant genetically modified crops can and are contributing to increased yields and agricultural growth in many developing countries and benefiting small-scale farmers. [256, 266]

Reports show that in the United Kingdom four genetically modified food materials have gained full approval and are in commercial use: cheese produced with genetically modified chymosin, tomato paste produced from slow softening tomatoes, and genetically modified soya and maize. Many other genetically modified products have cleared parts of the UK approval system (for example, clearance for food safety but awaiting environmental clearance for agricultural scale production). These include oil from oilseed rape, starch and oil from maize, oil from cotton, chicory, a variety of slow softening tomato intended to be eaten fresh, and riboflavin from a microbe. In addition, two other products granted full approval but have not been developed to full commercial scale are, genetically modified brewers' yeast and bakers' yeast. [255, 263]

Other statistics show that commercialization of biotech crops on a global basis is rapidly gaining ground. The ISAAA report shows that in 2008, the area dedicated to biotech crop cultivation grew by 9.4% above previous records (12.3 million hectares) reaching 12.5 million hectares, all over the world. Reportedly, this is the third highest area increase in the past five years, thus confirming the popularity of the crops. [255]

According to figures published by ISAAA (The International Service for the Acquisition of Agri-biotech Applications), the number of countries planting biotechnology crops has increased to 25, of which 12 are developing countries. South Africa is ranked eighth in terms of the acreage under GMO crops. In India, reports indicate that growers of biotech crops recorded an increase in income of up to $250 or more per hectare; increasing farmer incentives nationally from $840 million to $1.7 billion in the year 2007. As a testimony to strong farmer confidence in these new crop varieties, 9 out of 10 Indian farmers replant biotech cotton year after year.

In the US, between 50 and 70 per cent of the maize grown is genetically modified.

Other statistics show that China alone reported planting 250,000 insect-resistant biotech popular trees; that can contribute to reforestation efforts.

Australia is currently field-testing drought-tolerant wheat and two of its States recently lifted a four-year ban on biotech canola oil. In addition to planting more biotechnology hectares, an ISAAA report highlights that farmers are quickly adopting crop varieties with more than one biotechnology trait.

Sadly enough, Africa continues to lag behind in the adoption of biotech crops. To the continent of Africa, crop genetic engineering remains an enigma. Politics has unfairly been infused into this debate. Survey studies have indicated that most African countries are still dragging their feet on the issue of GMOs due to lack of basics like enacting facilitative biosafety laws to govern research, development and trade in biotech crops. The absence of legislation instruments intended to harmonize biotechnology operations has continued to stand in the way of existing initiatives that aim at promoting biotech crops. The lack of decisive measures has resulted in the slow progress made by many African states, in commercializing the GMO commodities. Thus, as the rest of the world angles itself to share the spoils of modern agricultural technology, African countries, with the exception of

South Africa that commercialized GM crops several years ago (and lately, Burkina Faso, Egypt, Cameroon, Tanzania, Malawi, Mauritius, Namibia, Mali, Zimbabwe, and Kenya), are still haggling on whether or not to admit biotech crops to their farms.

In South Africa, the GMO Act was passed by Parliament in May 1997 and the Regulations published in November 1999. The Executive Council, Registrar and Advisory Committee have all been appointed. Contravention of the Act can result in a fine or imprisonment of up to four years. [255]

In Kenya a legal path followed by those in support of commercialization of biotech crops has yielded fruits after the Bio-safety bill that sought to legalize biotech crops was recently endorsed and passed into law. A survey study carried out, [249] that Kenyan consumers are positive about the use of gene manipulation in food production; suggesting that there is a potential market for GM foods in Kenya. Reports show that even before the Bio-safety bill was passed into law in Kenya, some research organizations, including Kenya Agricultural Research Institute (KARI), Kenya Seed Company, and local universities have been spear-heading research on GMOs since 1998, on transgenic maize, sweet potato, cassava, cotton, and rinderpest vaccine. Monsanto, a US-based biotechnology firm and the Syngenta Foundation (associated with the Swiss-based biotechnology giant, Syngenta) and researchers at KARI have been experimenting on GM maize, cassava, sweet potatoes and cotton. There have also been open trials of genetically modified cotton on farmers' fields in some parts of Kenya including Mwea and Eastern Kenya. It is hoped that modalities of establishing a local regulatory body for GMOs, the National Bio-safety Authority (NBA) will be initiated to steer regulatory approvals and decision-making in moving the on-going field trials of GM insect-resistant cotton, maize and other products to the next level of commercialization.

In South Africa, cotton and yellow maize resistant to insects are commercially available; the latter accounts for about 2% of the total crop. Reports show that a number of GM crops are undergoing field trials in South Africa. The 1998 trials of planting insect resistant cotton in the Makatini region of KwaZulu Natal showed a significant decrease in the use of insecticides and an increase in yield of between 18 and 23%. The highest increase in yield was achieved by the small scale farmers involved in the trial and one farmer indicated having R30 000 in the bank which was not expected. [255, 257]

As with any other new technology, genetic engineering is not without adversaries, some of which even go as far as destroying experimental materials. This anti-science zealotry, [244, 256] and public hostility to modern biotechnology has been attributed to "lack of scientific literacy" [246, 256] and may impede human progress. [256]

One thing opponents of genetically modified organisms have overlooked is the fact that GMOs are the much awaited solution to improving food nutrition. [245] Today, an estimate of 852 million people in the developing countries suffers from hunger and malnutrition, and 1.3 billion people are affected by poverty. Thus undoubtedly, biotechnology can make a vital contribution to global food, feed and fiber security, and help feed the billions of poor people who constantly struggle for a better life. [250, 254, 256, 269] Studies show that already those farmers [worldwide] who began adopting biotech crops a few years ago are beginning to see socioeconomic advantages including increased earnings, and higher living standards, compared to their peers who haven't adopted the new crops. If we are to achieve the Millennium Development Goals (MDGs) (visualized by most developing countries) of reducing hunger and poverty to 50% by the year 2015, GMOs must play an even bigger role in the next decade. [255, 256]

LITERATURE CITED

[243]American Crop Protection Association, "Methods for Detection of GMO Grain in Commerce", September 2000.

[244]Borlaug, N.E. 2000. Ending world hunger. The promise of biotechnology and the threat of antiscience zealotry. Plant Physiol. 124:487–490.

[245]Bouis, H.E., B.A. Chassy, and J.O. Ochanda. 2003. Genetically modified food crops and their contribution to human nutrition and food quality. Trends Food Sci.-Technol. 14:191–209. Bouis, H.E., B.A. Chassy, and J.O. Ochanda. 2003.

[246]Bucchi, M., and F. Neresini. 2004. Why are people hostile to biotechnologies? Science 3004:1749.

[247]Buttel, F. H. (2002). The Adoption and diffusion of GM Crop Varieties: The "Green Revolution" in Global Perspective, 1996–2001." University of Wisconsin Madison, Program on Agricultural Technology Studies Paper Series, Paper no. 6.

[248]Cassman, K.G. (1999). Ecological intersification of cereal production systems; yield potential, soil quality, and precision agriculture. Proc. National Academies of Science USA 96: 5952–5959.

[249]Chege S. Kimenju and Hugo de Groote (2005). Consumer willingness to pay for GM food in Kenya.

[250]Conway, G., and G. Toenniessen. 2003. Science for African food security. Science 299:1187–1188.

[251]Diana Wong (2003). Genetically Modified Food Labelling and Research Services Division Legislative Council and

Secretariat. Available at:http://www.legco.gov.hk/yr0203/
english/sec/library/0203rp05e.pdf

[252]EFSA guidelines for the assessment of genetically modified
plants (May 2006).

[253]GAO (U. S. General Accounting Office). 2002. Genetically
modified foods: Experts view regimen of safety tests as
adequate, but FDA's evaluation process could be enhanced. United States General Accounting Office report
No. GAO-02-566. GAO Web Site.

[254]Huang, J., C. Pray, and S. Rozelle. 2002. Enhancing the crops
to feed the poor. Nature 418:678–684.

[255]International Service for the Acquisition of Agri- Biotech
Applications (ISAAA) Reports. Available at: http://www.
gmoafrica.org/2009/02/isaaa-reports

[256]James, C. 2003. Global status of commercialized transgenic
crops. ISAAA briefs no. 30. ISAAA, Ithaca, NY.

[257]Jauhar, P.P., and G.S. Khush. 2002. Importance of biotechnology in global food security. p. 107–128. In R. Lal et al. (ed.)
Food security and environmental quality in the developing world. CRC Press, Boca Raton, FL.

[258]Kitta, K. Mano, J. Furui, S. and Hino, A. (2009). The development of detection methods for the monitoring of GMO in
Japan. INRA/AFFRCS French-Japanese Seminar on Food
safety, novel food, and sustainable environment to promote animal and human health, INRA, Versailles – Paris,
France, June 8–10, 2009. Available at:http://www.nfri.
naro.afffrc.go.jp/research/katsudo/pdf/2009/6th_JFS_
Versailles_2009.pdf

[259]Liangzhi You, Mark W. Rosegrant, Cheng Fang, and Stanely
Wood (2005). Impact of Global Warming on on Chinese

Wheat Productivity. Environmental and Production Technology Division (EPT) Discussion: http://www.ifpri. org/divs/eptd/dp/papers/eptdp143.pdf

[260]Notification No. 1775 (June, 10, 2000). Ministry of Agriculture, Forestry and Fisheries, Tokyo 2000.

[261]Policy Planning Division, Department of Food Safety, Pharmaceutical and Food Safety Bureau, MHLW, "Mandatory Labeling of Genetically Modified Foods and Foods Containing Allergens", available at: http://www. mhlw.go.jp/english/topics/qa/gm-food/gm4.html.

[262]Qaim, M., and D. Zilberman. 2003. Yield effects of genetically modified crops in developing countries. Science 299:900–902.

[263]Radin, J.W. 2003. Lessons from a decade of genetically engineered crops. Agric. Res., January 2003, p. 2.

[264]Regulation (EC) 1829/2003 on GM food and feed – http:// www.food.gov.uk/gmfoods/gm/evaluating;

[265]Rosegrant, M.W. Cline, S.A. (2003). Global food security: challenge and policies. Science 302, 1917–1920.

[266]Rosenzweig, C. Parry, M. (1994). Potential impact of climate change on world food supply. Nature 367, 133–138.

[267]Schubert D "Regulatory regimes for transgenic crops" (2005). *Nat Biotechnol* [23]: 785–78.

[268]Walker-Simmons, M.K. 2003. New USDA-ARS research in biotechnology risk management. p. 94–98. *In* A. Eaglesham et al (ed.) NABC Report 15, Agricultural Biotechnology: Science and Society at a Crossroad. Natl. Agric. Biotechnol. Council, Ithaca, NY.

[269]Wambugu, F.M. 2001. Modifying Africa: How biotechnology
can benefit the poor and hungry, a case study from Kenya.
Nairobi, Kenya.

[270]Zepeda, L. (2002). "Genetically Engineered Food Labeling:
Consumers, Policies and Trade", Aug. 2002.